SpringerBriefs in Applied Sciences and Technology

Mark J. Jackson · Michael P. Hitchiner

High Performance Grinding and Advanced Cutting Tools

 Springer

Mark J. Jackson
Northborough R&D Center
Saint-Gobain High Performance Materials
Northborough, MA, USA

Michael P. Hitchiner
Saint-Gobain Abrasives, Inc.
Romulus, MI, USA

ISSN 2191-530X ISSN 2191-5318 (electronic)
ISBN 978-1-4614-3115-2 ISBN 978-1-4614-3116-9 (eBook)
DOI 10.1007/978-1-4614-3116-9
Springer New York Heidelberg Dordrecht London

Library of Congress Control Number: 2012942845

Printed on acid-free paper

Springer is part of Springer Science+Business Media (www.springer.com)

Preface

The market for cutting tools worldwide is estimated to be worth $200 billion, of which, 3 % of the market is dominated by abrasive cutting tools. The development of high-performance grinding wheels has allowed significant increases in the advancement in surface grinding processes such as creep-feed, VIPER, and high-speed stroke grinding. This brief describes the advances made in vitrified grinding wheels and abrasive grains that have enabled high-performance grinding to take place.

The first chapter describes the characteristics of abrasive grains in terms of shape and classification of those parameters for measuring particle shape and how they are related to sharpness, wear, toughness, and cutting ability. The different types of grains, their chemistry, and manufacture are explained together with how they are selected for various grinding applications. The second chapter of the brief is focused on the minerals formed during the firing of ceramic bonding systems, the reactions between alumina and the bond, and the provision of two case studies that detail how interfacial compounds and quartz-containing bonding systems affect wear of grinding tools.

The brief is written for students of materials science and manufacturing technology as well as practicing scientists and engineers involved in the manufacturing industries. The brief is meant to be used as an aid to understanding how abrasive grains and bonding system chemistry affect grinding wheel wear and performance in vitrified products. The brief also provides an understanding of how grinding wheels are formulated and manufactured.

Northborough, MA, USA

Ann Arbor, MI, USA

Mark J. Jackson

Michael P. Hitchiner

Contents

Chapter 1
Abrasive Tools and Bonding Systems

1.1 Abrasive Grain Characteristics

1.1.1 Grain Shape

Grain shape makes an enormous impact on grain strength, grinding performance, and packing characteristics that impact wheel formulation and manufacture. Shape will affect the r term in the undeformed chip thickness calculation $t' = \{[V_w/(V_s Cr)] (d/De)^{1/2}\}^{1/2}$ where r is the ratio of undeformed chip width to chip depth. This in turn impacts grinding power, finish, and force/grit. Shape and size are interlinked especially for particles of indeterminate shape, i.e., not perfectly spherical, cubic, etc. For synthetic diamond particles, for example, there exists an infinite combination of particle shapes derived from the transition between octahedral and cubic forms, as shown in Fig. 1.1. Moreover, crystal imperfections, and polycrystalline particles, further add to the variety of diamond forms.

A blocky round grain will in general be far stronger than an angular, sharp-cornered grain. Quantifying exactly what "blocky" and "angular" mean, and defining the characteristics key to shape have been the sources of considerable study both for grinding performance and batch to batch quality control of wheel manufacture. A variety of parameters for describing the shape of particle projections, classified according to the salient feature of the measurement, is shown in Table 1.1. The italicized parameters are those considered to be of most interest in a study by De Pellegrin et al. [1, 2], whereby the shape characteristics of diamond grain were studied in relation to grinding performance using optical microscopy and image analysis.

Two key diametrical dimensions are the major and minor diameters, d_a and d_b, respectively, as illustrated in Fig. 1.2, provide a fundamental measure of particle size. It may be noted that, depending on the orientation of the caliper lines relative to the given projection, many different diameters may be obtained. Although size is

M.J. Jackson and M.P. Hitchiner, *High Performance Grinding and Advanced Cutting Tools*, SpringerBriefs in Applied Sciences and Technology, DOI 10.1007/978-1-4614-3116-9_1, © Springer Science+Business Media New York 2013

Fig. 1.1 Diamond particle
shape map

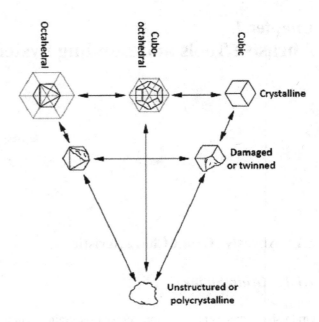

Table 1.1 Classification of parameters for measuring particle shape

Span	Area	Perimeter	Ratio	Average	Composite
Major diameter	*Projection area*	Silhouette perimeter	*Aspect ratio*	Average radius	Spike parameter
Minor diameter	Area moment	Perimeter moment	*Convexity*	Average diameter	Fractal dimension
Minimum diameter	Hull area	Hull perimeter	Circularity	Mean angle	*Sharpness*
Feret diameter	Equivalent radius (circle area based)	Equivalent radius (circle perimeter based)	Shape factor	Average curvature	Fourier/ Radance
	Equivalent ellipse		Ellipticity		Roughness Angularity

an important feature, it is shape that governs particle abrasiveness. Algebraic combinations of linear dimensions provide measures of shape.

Examples of such calculated values from this projection include:

1. *Aspect ratio*, the ratio of the major to minor diameter d_a/d_b. A useful parameter to describe grain elongation.
2. *Projection area*, the area enclosed by the boundary of its projection. It is an indirect measure of size and bulk of the particle. It is important as part of the calculation of grain convexity.
3. *Convexity* is a characteristic that strongly relates to the strength of the grain and its abrasive potential. A grain is convex if an idealized elastic membrane stretched

Fig. 1.2 Dimensional definitions for a grain 2D projection

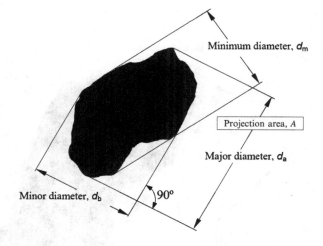

Minimum diameter, d_m

Projection area, A

Major diameter, d_a

Minor diameter, d_b 90°

across its projection leaves no space between itself and the grain's surface. The degree of convexity correlates with lower mechanical integrity but higher abrasive aggressiveness with the grain being, on average, less blocky. Convexity also correlates with the notion of grain irregularity.

Convexity C as a parameter is defined as

$$C = \frac{A_f + A_p}{A_p},$$

where A_p = the projected area of the grain, A_f = the filled area between the grain projection and the idealized elastic membrane stretched across the projection (Fig. 1.3).

4. Grain "sharpness" is a parameter that has been developed specifically for the characterization of abrasive grains based on chip formation modeling whereby the abrasive rate is governed by the degree of penetration into the workpiece, as illustrated in Fig. 1.4.

The functional relationship between the two orthogonal areas, Ω and Λ, is known as the "groove function," see Fig. 1.5. Averaging applied to numerous particle projections, and to the different particle orientations, yields the "average groove function" [1]. This function embodies the abrasive characteristics of an agglomeration of particles as might be found in a grinding wheel or coated abrasive paper.

From the groove function it may be seen that particle abrasiveness depends not only on the shape of individual particles but also on the complex distribution of the particles that constitute an abrasive surface [3]. The groove function embodies far greater information than any single parameter. The convenience of a single parameter led to the synthesis of the groove function into the "sharpness" parameter. The sharpness parameter involves rescaling the axes of the groove function by the

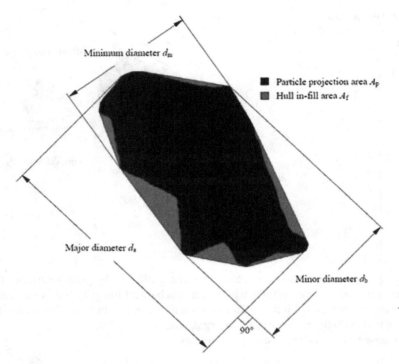

Fig. 1.3 Illustration of fill and projection areas for a grain 2D projection

Fig. 1.4 Orthogonal grain
projections relative to
penetration depth

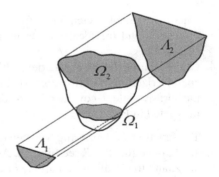

average projection area, A, of the particle sample. In this way, the groove function becomes non-dimensional—in terms of Ω_A and Λ_A—and particle samples of nominally different size can be compared solely in terms of their abrasiveness, which is governed by the slope of the curve. Sharper and more abrasive particles exhibit overall steeper slopes, and on this basis, integration of the dimensionless average groove function with respect to Ω_A, between zero and an arbitrary limit, L, yields the single parameter, sharpness (S), as illustrated in Fig. 1.6. The choice of L is governed

Fig. 1.5 Relationship between orthogonal areas provides the groove function

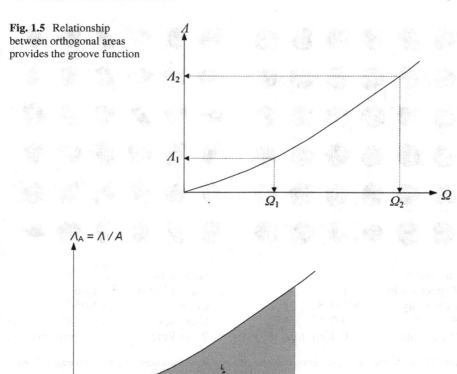

Fig. 1.6 Integration of the dimensionless average groove function to obtain the sharpness parameter, S

by the extent to which it is expected that the particles will penetrate the workpiece. A value of 0.3 was chosen because it represents a realistic average penetration during abrasion; i.e., the average indentation area, Ω, is 0.3 of the average projection area, A.

Having developed the sharpness parameter, De Pellegrin et al. [1] went on to compare it to convexity and aspect ratio for six different types of commercially available diamond grains (Fig. 1.7). The 2D projections of the two extreme-most samples (blocky → angular) are shown in Fig. 1.4. The difference in shape is visually apparent, but the strength of the sharpness technique is that it is able to quantify much subtler differences in particle abrasiveness. Wear rate measurements were made for each of the six samples. Wear tests consisted of abrading polyurethane blocks with constant load and grinding wheels specially fabricated from the candidate particles. A 0.99 correlation coefficient was observed between wear rate and sharpness, and a 0.98 correlation coefficient with convexity. Aspect ratio only managed a modest 0.82. While sharpness correctly ordered all particle types in

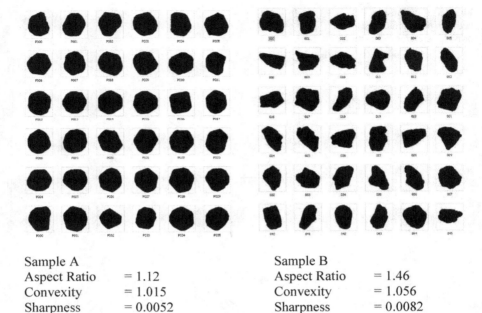

Sample A

Aspect Ratio	= 1.12
Convexity	= 1.015
Sharpness	= 0.0052
Wear Rate	= 0.4µm/min

Sample B

Aspect Ratio	= 1.46
Convexity	= 1.056
Sharpness	= 0.0082
Wear Rate	= 12.2µm/min

Fig. 1.7 Diamond grain projection sample extremes with associated shape characteristics and wear rates [1]

terms of wear rate, it was surprising to find that convexity offered almost the same performance as a predictor of wear rate, as illustrated in Fig. 1.8.

This interesting result suggests that convexity, a much more convenient parameter to calculate than sharpness, might be imbued with the qualities that make it a sound predictor of particle abrasiveness. This hypothesis needs further testing and verification. (For further detail on the various parameters discussed above see references [1, 4–10].)

Particle shape is fundamental to the performance of abrasives. However, the mechanical and thermal integrity of the grains can be just as important when grinding tenacious materials, as treated in the next section.

1.1.2   Attritious Wear Factors

Attritious wear is the wear that occurs "atom by atom" by interactions of the grain with the workpiece. These interactions are both physical and chemical, and can be quite complex. They can involve mechanical fracture (abrasion) at the microscopic scale and plastic deformation. Heat from friction and chip formation can lead to localized diffusion, chemical degradation and decomposition of the grain, and even

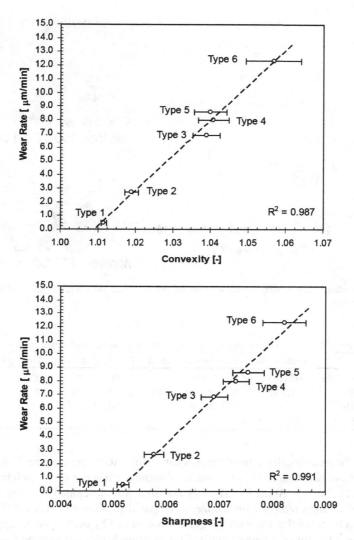

Fig. 1.8 The correlation of convexity and sharpness with wear rate experiments

melting. The clean surfaces exposed by the creation of a chip are highly reactive and can drive chemical reactions that would normally occur at much higher temperatures. Even the presence of oxygen in the atmosphere has a profound effect by neutralizing the clean steel chip surface. Grinding in a vacuum will generally lead to high levels of loading from metal to metal and grain to metal welding.

Hardness is the key factor in controlling attritious wear characterized by mechanical micro-fracture and plastic deformation. In general a grain has to be at least 20% harder than the workpiece to be suitable as an abrasive. Temperature plays an

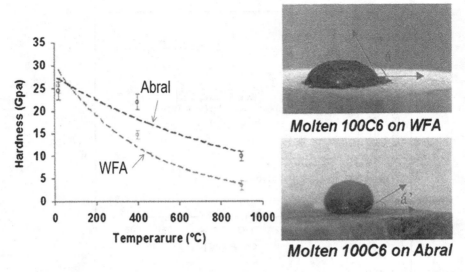

Fig. 1.9 Hardness and wetting characteristics of White Fused Alumina (WFA) and Al–O–N (Abral) abrasive

Table 1.2 Typical relative wear resistance values for major abrasive grain types during grinding

	Grinding alumina	Grinding steel	Grinding nickel	Grinding titanium
Diamond 9000 Hv	100,000	1,000	100	500
cBN 4500 Hv	1,000	10,000	5,000	100
Alox 1800 Hv	<1	5–10	10	1
SiC 2800 Hv	10	1–5	1	10

important factor as localized temperatures can easily exceed several hundred degree centigrade and hardness of abrasives such as alumina drops markedly with temperature. The Pechiney's Abral ™ Al–O–N abrasive was developed recently in part because of its high temperature hardness compared to alumina (Fig. 1.9) [11]. (It should also be noted that the Abral™ grain is not wetted by molten steel, suggesting an additional benefit of a lower tendency for welding leading to loading during the grind.)

The impact of hardness and other attritious wear controlling factors can be seen by the comparison of typical G-Ratio values for the major abrasive types such as diamond and cBN (cubic boron nitride), alumina and silicon carbide grinding various industrial workpieces (Table 1.2).

Grinding alumina the wear is essentially all mechanical for each abrasive type especially at low wheel speeds where heat generation is minimal. The effect in this case of hardness is very apparent. Diamond with its superior hardness provides a G-Ratio of typically 100 times greater than the second hardest abrasive, cBN, at slightly over half the hardness. Similarly, the G-Ratio for cBN is about 100 times higher than for silicon carbide with a similar proportional reduction in hardness. Alumina abrasive, with a hardness approaching that of the workpiece, gives a very poor G-Ratio and is in effect nonfunctional as an abrasive.

The situation with steel and nickel is very different. Diamond has a strong affinity for iron and even stronger affinity for nickel, both metals being true solvents of diamond at high temperature. cBN shows no such affinity. Therefore even though the mechanical wear should be in favor of diamond by 100:1 in fact the wear rate of diamond is at least ten times higher than for cBN. This is also seen in turning even at low speeds with relatively little heat generation. For example, a diamond tool turning mild steel wears 10,000 times faster than the same diamond tool turning aluminum–silica alloy of the same hardness. The reactions can be modified; for example, a diamond tool turning pure nickel wears 10,000 times faster than turning electroless nickel containing 10% phosphorus. The phosphorus appears to form a nonreactive nickel phosphide phase. Diamond is used to profile the faces of large electroless nickel coated mirrors to nanometer accuracies for satellite telescopes. The reactivity of diamond with iron can also be reduced in the absence of high temperatures by the presence of free carbon in, for example, pearlitic cast iron. Grinding cast iron at high speed and removal rates, cBN will out-perform diamond by two orders of magnitude. However when honing cast iron cylinder blocks at 2–3 m/s with the presence of free graphite, diamond will out-perform cBN by 10:1. If the carbon is present as carbide rather than graphite, however, the performance of diamond and cBN is reversed. Diamond is also used to burnish hardened steel and, with minimal heat and no clean metal created to make contact with the diamond, the life of the tool is extraordinarily long. Silicon carbide also shows high attritious wear grinding ferrous materials due to chemical and diffusion reactions with the workpiece to form carbides and silicides. Even aluminum oxide will react with iron in the presence of oxygen at high enough temperatures although this is generally not a significant enough factor to impact the economics of wheel life. Komanduri and Shaw [12] has identified chemical and diffusion wear processes in the grinding of ferrous alloys with SiC, in particular, the formation of carbides (Ni_3C and Co_3C) grinding superalloys.

Finally, the grinding titanium is probably the most complicated with numerous wear mechanisms including chemical, diffusion, adhesion, and mechanical impacting the economics of wheel life versus grain cost. Titanium has a very low thermal diffusivity factor ($\kappa c \rho$), approximately a quarter that of steel, such that for comparable grinding condition the temperatures of the workpiece/wheel interface can be doubled [109]. Titanium shows a chemical affinity for SiC and alumina grain with reactions of the form [13]:

Silicon Carbide:

$$SiC + O_2 \rightarrow SiO_2 + C$$
$$SiC + 2O_2 \rightarrow SiO_2 + CO_2$$
$$Ti + SiO_2 + O_2 \rightarrow TiO_2 + SiO_2$$
$$Ti + SiC + O_2 \rightarrow TiC + SiO_2$$

Diamond also shows some affinity with titanium and will form the TiC phase at high enough temperatures but the reaction rate is not on the same order of magnitude as for the true solvent metals (e.g., Fe, Ni) with carbon. Instead we might expect some impact from high adhesion forces. cBN, or more precisely the oxide layer formed on the surface of the cBN grains from reaction with oxygen in the air or water in the coolant, is also somewhat reactive with titanium. What studies have been made of wear mechanisms of cBN on titanium have generally been in turning. Here, the primary chemical wear mechanism was with the binder between the cBN grains in compact tools. With recently developed binderless cBN tools the attritious wear mechanisms were primarily a mixture of mechanical, adhesion, and diffusion with no dominating chemical wear factor [14]. Grinding studies with electro-plated and bonded wheels in general indicate diamond out-performs cBN [110, 111] although this is not yet definitive [112]. Kugemai et al. [13] showed that, when grinding titanium, diamond abrasive gave the highest G-Ratio, lowest grinding forces, and lowest grinding temperatures of all the standard abrasive types including alumina–zirconas while alumina-based grains gave the worst. High pressure coolant was a significant aid further indicating the importance of adhesion mechanisms. Kumar [15] found that a medium toughness diamond abrasive could grind titanium alloys more efficiently than cBN or SiC. Nevertheless, from the standpoint of economics, SiC is still the abrasive of choice in industry based in part on abrasive cost but more on ease of use, especially dressing of forms critical for aerospace and medical applications, while being able to generate acceptable workpiece quality. This will become an area of increased research both with the commercialization of titanium intermetallics such as γ-TiAl for aerospace, power generation, and diesel turbochargers, and with the rising cost of SiC abrasive.

1.1.3 Grain Fracture Toughness

Whereas hardness provides a measure for the tendency of grain to wear by attrition on the atomic scale, fracture toughness (or it's inverse term "friability") provides a measure for the loss of abrasive due to breakdown by fracturing or splintering of the grain typically at the micron level (micro-fracture) or greater (macro-fracture). The degree of fracture is in large degree dependent on grain properties such as crystal size and morphology, impurities, inclusions and preexisting cracks, and shape. It is also very dependent on the level and nature of the forces applied to the grain during grinding and from factors in the grinding environment such as thermal shock from coolant. Attritious wear leads to the creation of wear flats that dramatically increases the force exerted on the grain and in turn leads to increased levels of fracture.

Fracture toughness, particularly of superabrasive grain, is most commonly evaluated by a vibration–impact test. A grain sample of a known particle size distribution is placed in a tube with steel ball bearings and shaken with a fixed amplitude and frequency for a given length of time. The grain particle size distribution is then re-measured to assess the level of breakdown. The grain is either measured as received

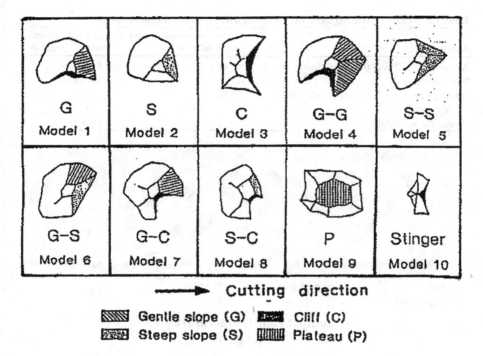

Cutting direction

Gentle slope (G)   Cliff (C)
Steep slope (S)   Plateau (P)

Fig. 1.10 Primary edge models for the morphology of fractured grains [16]

to give a toughness index (TI) value; or after processing at high temperatures, typical of those seen in wheel manufacturing process or use, to give a thermal toughness index (TTI) value. The high temperature processing can occur either in vacuum, or in the same atmosphere as that used in wheel manufacturing, or even after mixing with wheel bond, e.g., glass or vitrified frit, which is dissolved with HF acid subsequent to heat treatment. In general the TTI will be less than the TI as temperature causes the expansion of inclusions, reactions with the atmosphere, and infiltration of surface flaws with glass bond. Where the grain has previously been through a significant degree of crush processing, especially for fused alumina grain, high temperature calcining can actually increase the TTI by annealing existing cracks. Crush strength measurements are also made on single grains.

Hagiwara et al. [16] have developed a method of evaluating the grain strength from fly cut measurements using single grains. They evaluated the grain strengths in terms of a fracture onset probability but also categorized the shape of the fractured grains. They proposed ten primary edge models for the morphology of fractured grains (Fig. 1.10). Their study reports measurements made on populations of grains using examples of alumina and SiC abrasives.

Probability of survival of a grain is given by $P_t = 1 - e^{-\gamma t}$ where γ is the fracture onset coefficient. Values for γ and primary fracture modes are given in Table 1.3.

Table 1.3 Fracture onset and mode of fracture data from fly cut measurements [16]

Abrasive type	Fracture onset coefficient (mean value γ_{mean})	Fracture modes (most common)
Alumina/2.3% titania (brown alumina)	0.046	1,10,5
White alumina	0.174	10,3,8
Single grain, temperature annealed, alumina	0.142	1,4,5
Black silicon carbide	0.407	10,2
Green silicon carbide	0.317	10,3,5

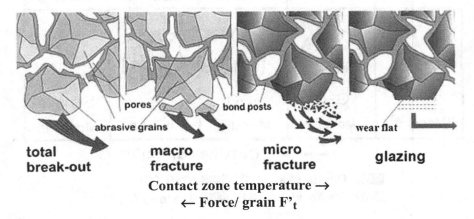

Fig. 1.11 Grain/bond breakdown modes in grinding wheels (after a drawing by Rappold)

The technique readily distinguishes between alumina and SiC, showing the highly friable nature of the latter, as well as more subtle differences between various grades within a grain family. Grain toughness must be matched to both the wheel bond characteristics and the grinding conditions (Fig. 1.11).

Ideally the grain should fracture creating the loss of relatively fine particles typically at the micron or submicron level; a process termed "micro-fracturing." The remaining portion of the grain should remain sharp and able to cut. If the grain is too tough relative to the bond holding it, or the grinding force/grain is extremely high, then the grain can undergo total break-out or loss without doing any useful work. If the bond is strong enough to hold the grain but there are high grinding forces/grain, and/or the grain crystallite size is large, then the fracture is often more one of coarse loss of grain by "macro-fracturing" still without the full amount of possible useful work being obtained. On the other hand if the grain is much weaker than the bond and/or prone to high attritious wear due to mechanical, heat, or chemical wear, then "glazing" occurs resulting in the creation of wear flats, high grinding forces, and increased interface temperature. Higher forces will lead in turn to more fracture.

The ideal stable state for wheel wear is a limited amount of attritious wear controlled by micro-fracture. The maximum amount of wear flat area is set by the onset of thermal damage. For ferrous materials this limit is about 1–2% of the wheel surface when using alumina or SiC abrasives, and about 4–5% when using cBN or

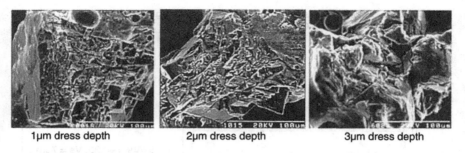

| 1μm dress depth | 2μm dress depth | 3μm dress depth |

Fig. 1.12 Micro-to-macro-fracture trend rotary diamond dressing cBN at increasing dress depth

diamond abrasives due to their higher thermal diffusivity or ability to remove heat from the grind zone.

Fracture behavior is also important in terms of the grains reaction to impact during dressing. This is especially true of cBN in vitrified bonds for high production grinding. Ishikawa et al. [17] reported a study of dressing vitrified bonded wheels containing coarse grade (80#) GE 1 abrasive using rotary diamond disk dress rolls.

It was determined that at a modest crush ratio of +0.2 there was a definite shift from predominantly a micro-fracture regime at a depth of dress of 1 μm to a macro-fracture regime at a depth of dress of 3 μm (Fig. 1.12). Changes to dress depth of as small as 0.5 μm had a significant effect on power and finish. As the crush ratio was increased from +0.2 to +0.8 the level of macro-fracture increased dramatically to dominate the process. These results are specific to a particularly friable grade and size of cBN typical of that used in resin and weaker vitrified bonds. It is therefore to be expected that a tougher grade of cBN or a finer grit size would require either a higher crush ratio and/or deeper depth of cut to achieve the same degree of fracture. Evidence for this is suggested in the work by Takaga et al. [18] who found that when dressing 80# GE 500 abrasive at a depth of dress of 5 μm micro-fracture dominated even at a crush ratio of +0.5 but at a crush ratio of +0.9 was replaced by macro-fracture. GE 500 is a particularly tough grade used primarily in single layer plated applications.

The control of the level of micro-fracture and the resulting wheel surface morphology is especially critical to dressing of vitrified cBN wheels with rotary diamond dressers for high production grinding. A major challenge with vitrified cBN is the initial rapid changes in grinding characteristics immediately after dressing especially for relatively weak systems or burn sensitive grind operations. The effect is illustrated in Fig. 1.13 with examples by Jakobuss et al. [19] of normal force changes for different dress crush ratios and wheel speeds. The problem is in the first perhaps 5% of possible grinding between dress cycles. The effect is also seen with conventional abrasives to some extent but is fleeting being essentially completed before the first workpiece component has been finished ground. For a cBN ground process with several hundred workpieces being ground between dresses this rapid

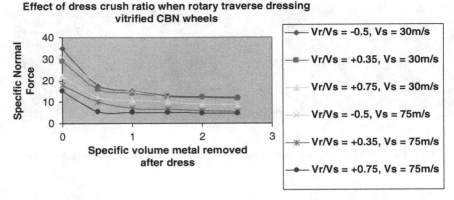

Fig. 1.13 Influence of crush ratio on normal grinding force when rotary diamond dressing of vitrified cBN wheels

change in grinding force can lead to taper and burn issues or the need for special programming routines to reduce feed rates to compensate.

Fujimoto et al. [20] made a detailed study of the wear of grain in the surface of a vitrified cBN wheel by means of three-dimensional multiprobe SEM (scanning electron microscope) and fractal dimension analysis. Grinding conditions along with force, finish, and wear data are given in Fig. 1.14. They identified three stages in a complicated sequence of events as the wheel wore. Immediately after dressing they observed a break-in period with higher wear and the characteristics drop in grinding forces. Observations of individual grains shown in Fig. 1.15a–e found that this drop was associated with a loss of unstable grain edges and the formation of new sharp cutting edges. Analysis of three-dimensional profiles reveals that the cutting edge density is reduced (Fig. 1.16) affecting the grain density in the wheel surface to a depth of at least 20 μm. After the initial break-in there is a steady state condition in terms of wear rate and change in finish. However, this steady state can be divided into two regions. In the first region there is micro-fracturing of the grains and signs of attritious wear building on the cutting edges. Interestingly the grinding forces climb slightly even though the finish is climbing too, indicating major changes in cutting edge density and grain edge shape. The second region begins after about 15 μm of wheel wear; at this point the wear flats on the grains are pronounced and macro-fracturing of the grains dominates. The grinding forces now remain constant.

The concept of micro-truing and macro-truing in fact has far greater for range and control in the latest conventional abrasive technology based on sol- or seeded-gel sintering processes or, most recently, agglomeration techniques.

Interestingly, in a high production manufacturing environment using 80# cBN grain, the amount of wear between dresses is usually limited to about 10–15 μm in order to keep the process under control over repeated dress cycles. Earlier research by Yonekura et al. [21] and Mindek [22] described a surface affected layer termed "Tsukidashiryo," or "Active Surface Roughness" generated by the dressing and

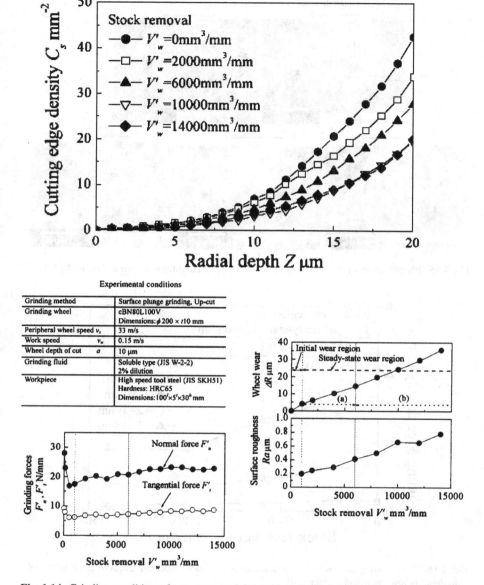

Fig. 1.14 Grinding conditions, force, wear, and finish from study on cBN wheel wear [20]

grinding processes, and varying in depth from a few microns to over 30. After the initial dress and grind cycles the surface is conditioned such that the dress amount of the second dress cycle is now critical. Over-dressing, that is removal of total depths of >20 μm, will result in a closed wheel similar to the first dress; under-dressing, that is removal of total depths of <5 μm (for 80# grain size), will result in

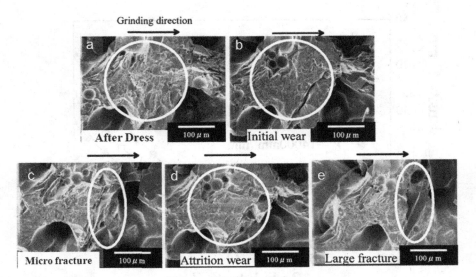

Fig. 1.15 Photographic study of cBN grain breakdown for conditions shown in Fig. 1.14 [20]

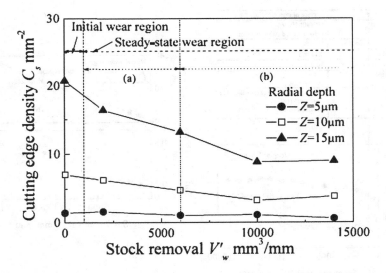

Fig. 1.16 Grain cutting edge density variation with workpiece material removed after dress and with depth below the wheel surface [20]

fewer parts/dress as the cutting edge density is now too low. Much of the optimization involved with vitrified cBN processes is in the selection of the appropriate dress depth per pass to control the level of micro-fracture and the total dress depth to control the cutting edge density. This in turns limits the amount of the magnitude of the break-in period while maintaining an optimum parts/dress.

The application of micro-truing to grain structure is not limited to cBN. A growing awareness of the benefits of controlling fracture at the micro-level has led in recent years to a whole new family of "engineered" alumina abrasive grain structures.

1.2 Silicon Carbide

Silicon carbide (SiC) was the first of the synthetic abrasives that ushered in twentieth century manufacturing. It was first synthesized in commercial quantities around 1891 by Dr. Edward G Acheson [23] who gave it the trade name "Carborundum," and was initially produced in only small quantities and sold as a substitute for diamond powder for lapping precious stones at $880/lb (at 1891 dollar value!). With process optimization the price fell precipitously $0.10/lb in 1938. Today (2011), the price is about $0.80/lb. The heart of the process is the Acheson resistance heating furnace, an adaption of the Cowles electric batch smelting furnace patented just a few years previously in 1885, in which a mixture of quartz silica sand and petroleum coke is reacted at a temperature of around 2,400°C [24]. The overall reaction is described by the carbothermic reduction equation:

$$SiO_2 + 3C \rightarrow SiC + 2CO$$

The furnace is prepared by placing a large carbon resistor rod on a horizontal bed or trough of raw materials to which a heavy current is applied. The raw material also includes sawdust to add porosity to help release the CO and salt to remove iron impurities. The whole process takes from 36 h to 10 days and yields typically 10–50 tonnes of product. From the time it is formed the SiC remains a solid as no melting occurs (SiC sublimates at 2,700°C).

Pure SiC is colorless. Two grades of SiC are produced for abrasive applications— "green" and "black." Green SiC is the purer produced from a virgin mix of sand and coke; black SiC is produced from recycled feed including amorphous SiC from previous furnacing cycles; the black coloration comes from iron impurities. Green and black products are also sorted in terms of distance from the carbon rod, with some green material being obtained closer to the electrode even with recycled feed.

Figure 1.17 shows photographs of a Saint-Gobain SiC plant in Norway. The first photograph is a view of the filled bed of an Acheson furnace, the second photograph shows the product after it has been removed from the furnace with the carbon rod still embedded in the center of the ingot. After withdrawal of the rod and removal of the surrounding amorphous SiC the remaining mass is 98% SiC to be further processed.

The Acheson process has remained essentially unchanged for many decades. As such the primary driver for the location of manufacturing has been cheap, easily available electric power—most commonly hydro-electric power. The original Acheson furnaces were driven by power from Niagara Falls although North

American production is now severely challenged by operating costs; global manufacturing is now dominated by China with almost half the market. Other countries with significant grain production include Brazil, Russia, and Vietnam. More recently Czech, Spain, and even Bhutan have come on line although much of this product is for other applications than abrasives. There has been a large upsurge in interest in SiC for applications such as tank and body armor, heat resistant bodies for kiln ware, high temperature electronic devices, aero-engine components, and wire saw applications for electronics spurring research into both manufacturing processes and SiC material properties. Silicon carbide is the hardest of the conventional abrasives with a Knoop hardness of 2,500 and a surface Knoop microhardness of 2,900–3,100 kg mm^2 at room temperature. Microhardness falls off with temperature as shown in Fig. 1.18 [25, 26].

Green silicon carbide is the higher purity silicon carbide manufactured with typically >98.5% of SiC. The crystal type is alpha phase silicon carbide in the form of hexagonally shaped platelets. Black silicon carbide is of a lower purity (95–98%) and consists of the alpha phase with both hexagonal and rhombohedral forms. The green is the slightly harder but more friable and angular. For this reason green silicon carbide is used for grinding hard metals such as chilled cast iron rolls, titanium, and metal and ceramic cutting materials. Black silicon carbide is used more for grinding of soft nonferrous metals and nonmetallics like rubber, wood, ceramics,

Fig. 1.18 Hardness
variation with temperature
for SiC

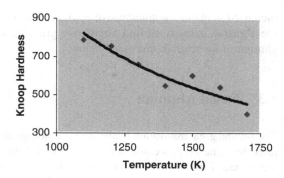

Fig. 1.19 Examples of black and green SiC

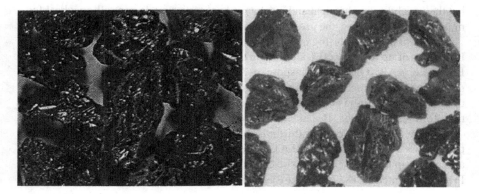

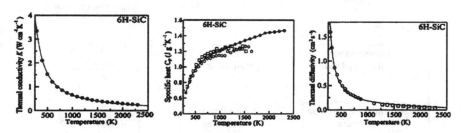

Fig. 1.20 Thermal properties—κ, C_p, and $(\kappa C_p)^{1/2}$ of SiC

and glass. Both SiC grades are more friable than fused alumina grain (Fig. 1.19).
SiC does show reactivity or solubility with iron, limiting its use in grinding ferrous
materials. It is also susceptible to oxidation at higher temperatures. Thermal properties are shown in Fig. 1.20 [27].

The demand for SiC is such that the price of abrasive grain has risen steeply in
the last few years on a trend that is expected to continue due to energy costs and

increased demand for nonabrasive applications. This presents a major challenge to the abrasive industry to find alternative grinding solutions such as agglomerated aluminum oxide grain discussed below.

1.3 Fused Alumina

The second great innovation in abrasive technology at the turn of the twentieth century, after the Acheson process for SiC synthesis, was the invention of the Higgins electric arc furnace for the production of electro-fused alumina (or "Alundum") by Aldus C. Higgins of the Norton Company in 1904 [28]. Prior to this, wheel makers had used naturally occurring aluminum oxide in the forms of the minerals emery and corundum but the variability in chemical and mechanical properties made controlling wheel formulations difficult. Now the use of emery abrasive is limited to coated paper.

Raw bauxite is the starting material for all fused aluminum oxide grain (Fig. 1.21). It consists of up to about 60% alumina in the form of the minerals gibbsite $Al(OH)_3$, Boehmite γ-AlO(OH), and diaspore α-AlO(OH), together with the iron oxides goethite and hematite, the clay mineral kaolinite, and small amounts of anatase and titania, TiO_2. Australia is the largest producer with almost a third of world production, followed by China, Brazil, Guinea, and Jamaica [29].

Bauxite along with coke and iron is the direct feed material for fusion to create the Brown Fused Alumina (BFA), a family of abrasives containing controlled amount of up to 4% titania. Bauxite can also be purified prior to fusion by the Bayer Process invented in 1887 by Karl Bayer in Russia. In this case, bauxite is heated in

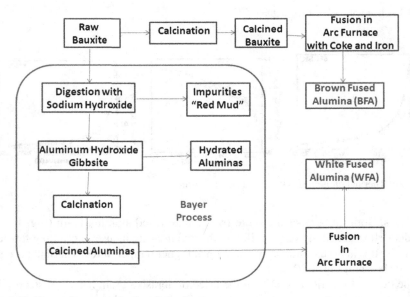

Fig. 1.21 Processing routes for fused alumina

Fused Higgins shell in water cooling   Electrode arm assembly

5 MVA Higgins EAF in operation   Electrode of Higgins EAF under power

Fig. 1.22 Examples of Higgins furnaces and equipment details (Courtesy Whiting Equipment, Canada Inc.)

a pressure vessel with sodium hydroxide solution at 150–200°C. After separation of the iron-based residue (red mud) by filtering, pure gibbsite is precipitated by cooling the liquid and seeding with fine-grained aluminum hydroxide. Gibbsite is then converted into aluminum oxide by calcining. The Bayer process removes almost all of the natural impurities present in the raw bauxite, but leaves behind 0.1–0.4% soda (Na_2O) in the purified calcined alumina. This is the feed for the production of White Fused Alumina (WFA) and its family of abrasives. The Bayer Process increases the cost of WFA feed material by about a factor of 5 compared with bauxite feed for BFA. A Higgins electric arc furnace consists of a thin steel or aluminum shell on a heavy metal hearth (Fig. 1.22). A wall of water running over the outside of the shell cools it sufficiently to maintain the shell integrity in combination with a thin layer of aluminum oxide that forms on the inside due to its extremely poor thermal conductivity. Steel was historically the normal shell material, as it has a relatively high melting point, but aluminum is now preferred especially for WFA fusions to prevent discoloration from iron rust contamination. Feed material is poured into the bottom of the furnace and a carbon starter rod laid on it. Two or three large vertical carbon rods are then brought down to touch and a heavy current applied. The starter rod is rapidly consumed but the heat generated melts the bauxite, which then becomes an electrolyte. Feed material is added continually over the next several hours to build up the volume of melt to as much as 20 tonnes. Current flow is controlled by adjusting the height of the electrodes which are eventually consumed in the process. The reaction conditions of BFA fusion as a result of the added coke which reacts with the oxygen in the impurities to produce carbon monoxide, reduces the silica to silicon and iron oxide to iron which combine along with the added iron to form a heavy, highly fluid, ferrosilicon phase. Silica is also lost as fume due to the high temperatures. In addition titania levels may be adjusted by reduction to titanium that precipitates out with the ferrosilicon. A typical fusion of 30 tonnes may take around 20 h to completely fill and melt the contents of the furnace pot, cooling time may be up to 4 days. The cooling is very directional; the insulating outer layer of the melt is quenched next to the pot and highly microcrystalline. There is then a large crystalline, dendritic, growth region in a radial direction with solidification towards the center of the pot as heat flows from the center out. The pot has a high profile with an aspect ratio of about 1:1. Impurities will concentrate in the liquid phase in the center and towards the bottom of the forming ingot. After cooling the ingot must be broken up and hand sorted to remove the primary concentration of impurities. Additional iron and ferrosilicon are subsequently removed with magnetic separators during crushing (Fig. 1.23).

The reaction conditions for a WFA fusion are in general not considered reducing in that the only carbon present is from the electric arc and starter rods. The biggest concern is the conversion of the residual soda from the Bayer process feed into Sodium β-Alumina which crystallizes as soft hexagonal plates in the alumina. As Sodium β-Alumina has a lower melting point than alumina it will again concentrate in the in portions of the ingot that solidify last. The Higgins furnace has evolved from an original design with a small capacity of 1–5 tonnes to furnaces today of up to 40 tonnes with pot diameters of up to 3.5 m (12 ft) and a power supply of up to 4 MVA. It requires 2.2 MVA h to produce 1 tonne of BFA and 1.5 MVA h to produce 1 tonne of WFA. However, with increasing capacity and efficiency demands there has been a move towards even larger tilting furnaces up to 6 m in diameter that can pour the molten alumina into pots with water cooled hearths. These furnaces use a power supply of as much as 10 MVA or even greater and can pour up to 24 tonnes every 4 h whilst maintaining a more consistent batch to batch chemistry.

Pour pot design for use with tilt furnaces can have a major impact on grain structure and chemistry (Fig. 1.24). For example, for WFA fusion pour into a high profile pot the cooling process and output is similar to a Higgins furnace, i.e., a large alumina crystallite size with dendritic growth and a very low sodium β alumina content after sorting. However pouring into a low profile pot (aspect ratio << 1) onto a cold hearth results in a much faster cooling rate, a fine crystallite alumina structure and a much more evenly dispersed—but higher—sodium β-alumina content. The high thermal gradient when cooling a WFA ingot in a deep profile pot causes crystallization

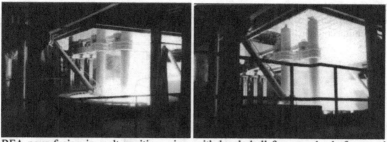

BFA pour fusion in melt position; view with level shell from work platform and pouring floor

Preparing to pour, electrodes up and feed chute clear. Pouring mold (on mold car) receiving molten BFA.

Molten BFA in mold after tilt pour.   Deep tilt pour of ferrosilicon in crucible

Fig. 1.23 Examples of electric arc tilt pour furnaces and operation (Courtesy Whiting Equipment, Canada Inc.)

of α-alumina in a dendritic habit made up of inter-grown rhombohedra extending along the thermal gradient. This type of crystal is the result of the edges of the rhombohedron growing much faster than the faces (Fig. 1.25). Abrasive grain made from this will tend to fracture in relatively large fragments along well-defined planes but be very self-sharpening.

Crystallization in low profile pots will show structures with less directional growth and that are finer in crystal size resulting in smaller fragments during grain fracture; the material is also about 10% softer from higher sodium β-alumina contamination. For BFA fusions in tilt pour furnaces about 25% of the furnace

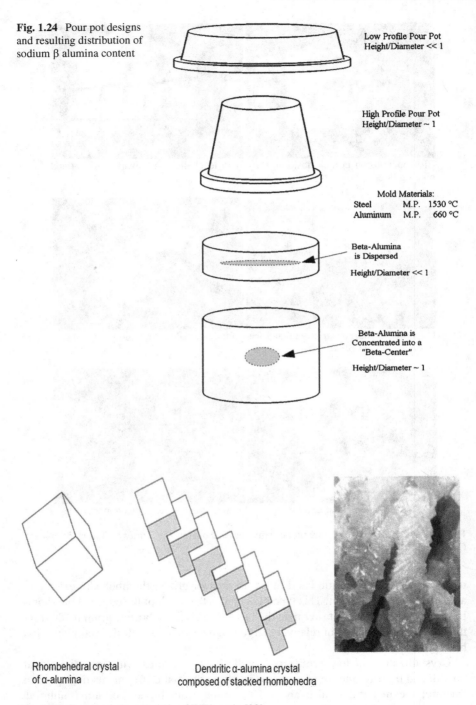

Fig. 1.24 Pour pot designs and resulting distribution of sodium β alumina content

Low Profile Pour Pot
Height/Diameter << 1

High Profile Pour Pot
Height/Diameter ~ 1

Mold Materials:
Steel   M.P.   1530 °C
Aluminum   M.P.   660 °C

Beta-Alumina
is Dispersed

Height/Diameter << 1

Beta-Alumina is
Concentrated into a
"Beta-Center"

Height/Diameter ~ 1

Rhombehedral crystal
of α-alumina

Dendritic α-alumina crystal
composed of stacked rhombohedra

Fig. 1.25 Structural characteristic of WFA grain [30]

content is poured each time while most of the ferrosilicon collects at the bottom where it can accumulate over many regular pours until it is removed in a "deep pour" into a rail cart with a bed of sand. The quality and consistency from large pour furnaces tends to be better because it is less expensive to monitor per tonne taking regular dip samples; in particular avoiding over-reduction of the titania. Pour molds tend to be low profile and the resulting grain a mixture of dendritic with finer equiaxial α-alumina together with some ferrosilicon inclusions. For further discussion on grain fusion and furnacing see Wolfe et al. [30], Lunghofer et al. [31], and Whiting Equipment Canada [32].

The fusion of "Alundum" began over a hundred years ago on the shores of the Great Lakes using cheap hydro-electric power from Niagara Falls. Today, 2012, only three plants remain in north America with just 5% of global capacity although fused alumina has been substituted to a significant degree with more recent ceramic technology and manufactured as discussed below. In the last 12 years China has increased its capacity for fused alumina, especially BFA, to over 60% USGS [33, 34] with integrated manufacturing close to the bauxite mines and the largest capacity tilt pour furnaces. Eastern Europe, India, South Korea, and South America also continue to increase their prominence in world markets. Low cost electrical power availability, furnace capacity, quality control, and cost of raw material sourcing should be the dominant factors influencing the relative competitive stance of each.

1.3.1 Grain Types

Properties of the grain depend both on the fusion process and chemistry but also on the subsequent comminution process. The ingot is initially split and sorted and then pre-crushed in steel jawed Barmac and beater crushers. Further crushing is produced by passing the material through roll crushers. All these processes are highly impact and will create major fractures resulting in a grain that is sharp edged, flawed, and anisotropic, typically like a sliver in shape. Subsequent processing in steel or rubber lined ball mills will have a tendency to reduce grain size by rounding of the grain edge. In this manner it is possible to control shape to a degree to have either angular or blocky forms from the same material. The grain availability can be divided into BFA- and WFA-based families

Brown fused aluminum oxide contains 2–4% titania which enhances toughness. This is still the most widely used abrasive in wheels to grind high-tensile-strength materials, and for rough grinding, deburring, and snagging, as well as to cut low-alloy, ferrous materials and is generally viewed as the "workhorse" of the industry. Brown Fused Alumina is a tough, sharp but blocky abrasive. Depending on the processing regimes the grain is typically about 50% single crystal and can be provided in high, medium, and low density based on shape packing characteristics. The grain may also be calcined after sizing to toughen it by annealing cracks generated in the crushing processes. The material is sometimes termed blue fired BFA as the grain changes color due to surface oxidation of impurities. Specialty coating such as

Fig. 1.26 Examples of (**a**) white, (**b**) pink, and (**c**) red-fused alumina grain

silane (for resin bonded wheels to resist coolant interactions) or red iron oxide (for resin and rubber bonded wheels to increase surface area) may also be applied.

Low titania ("light" or "semi-friable") brown fused aluminum oxide has 1–2% TiO_2 content and is used in bonded or coated applications that require an abrasive that is slightly tougher than white aluminum oxide. Reducing the titania content reduces the abrasive's toughness, but increases its friability. Light BFA is commonly used in depressed center wheels, cut -off wheels, and for surface and cylindrical grinding of heat sensitive metals, alloys, etc. where cool but fast cutting is required. The grain can be supplied with similar post and surface treatments as regular BFA.

White Fused Alumina is the standard multicrystalline WFA with sodium β-alumina contamination and is the most friable grain in the fused alumina family. It is considerably harder than BFA. The most common applications include the grinding of tool, high-speed and stainless steels (Fig. 1.26).

Single crystal White Fused Alumina is the single crystal grain that has been produced in deep pour fusion pots and separated from any sodium β-alumina contamination. This is the hardest and most brittle of the alumina family of grains used most commonly for grinding tool and very high alloy steels that are very sensitive to heat.

Fig. 1.27 Example of sintered extruded brown alumina grain (Courtesy Saint-Gobain Abrasives)

Pink alumina is WFA to which <0.5% chromium oxide has been added in the fusion process to produce a grain that is slightly tougher than regular WFA used for grinding unhardened high alloy steels (Fig. 1.26).

Ruby alumina is WFA to which 3% chromium oxide has been added to provide additional toughness over pink alumina (Fig. 1.26).

It can be inferred that there is a steady increase in toughness but reduction in hardness in the following order:

$$\text{Single crystal WFA} \rightarrow \text{WFA} \rightarrow \text{pink WFA} \rightarrow \text{ruby WFA}$$
$$\rightarrow \text{light BFA} \rightarrow \text{BFA} \rightarrow \text{Blue fire BFA}$$

In general the wheel maker will blend various grain types and sizes to combine the properties of each. In addition to chrome, other metal oxide additions have been investigated including vanadium and beryllium but not found to be commercially viable.

Sintered Alumina is a family of grains developed in the 1950s produced from *unfused* alumina. Several processes exist based on both raw bauxite and Bayer processed aluminas. The most common is to use a feed material of raw bauxite milled to <5 μm. The mix with binder is first extruded to produce rods which are cut into short cylinders or cones in the green state. They are then fired in rotary kilns at 1,350–1,500°C using natural impurities in the bauxite as sintering agents (Fig. 1.27) [35]. The resulting grain is extremely tough especially at the relatively large sizes the technology allowed to be produced (8# – 20#) and the material found great success, until the advent of alumina–zirconia grain, in billet conditioning and other rough grinding operations. It is still used as a blend component with alumina–zirconias especially in the grinding of stainless steels.

1.4   Alumina–Zirconia

Zirconia is a very high temperature refractory material, tougher but softer than alumina. It also has a higher melting point than alumina making fusion based on the Higgins furnace more demanding in terms of containment and control. Fortunately, from a manufacturing viewpoint (if not a grinding perspective), the thermal conductivity of zirconia like alumina is very low making the Higgins fusion route still possible. However, as can be seen the liquidus curve for the zirconia–alumina system (Fig. 1.28) [36], compositions of these two materials combined, with the zirconia contents kept under about 65%, have comparable or lower melting points to alumina alone. This allows for relative ease of control of a combined alumina–zirconia fusion process including pouring from a tilt furnace.

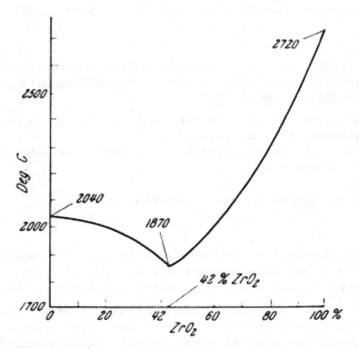

The liquidus curve for the system Zirconia-Alumina

Fig. 1.28   Phase diagram for the Zirconia–Alumina system

Interest in zirconia as a potential abrasive grain or component of grain has been evident from the literature since at least the mid-1950s, e.g., [37, 38], in part because a relatively pure form known as baddeleyite began being produced in quantity at that time as a by-product of heavy metal mining especially of uranium ores in countries such as Russia, Brazil, and South Africa. Another much commoner source of zirconia is as zircon (zirconium silicate, $ZrSiO_4$) found as sands in the USA, Australia, and Brazil. Zircon sand can be refined by fusing it with coke, iron, and lime until the silica is reduced and separates out as a denser, relatively low viscosity ferrosilicon liquid. Alumina–zirconias can be produced in a similar way by the addition of Bayer process alumina to the fusion. Zirconia has the interesting property that it is metastable in the tetragonal state at certain crystal sizes when held under constraint. For pure zirconia the grain size upper limit is about 0.1–0.3 m. With additions of small amounts of the alkali oxides CaO or MgO, or rare earth oxides such as Y_2O_3 or CeO_2, this limit can be raised into the micron range. Without constraint the tetragonal crystal converts to the monoclinic phase with a significant increase in volume of about 6%. If a crack from an active fracture intersects with a tetragonal crystal it releases the constraint but in the process the volume expansion to the monoclinic phase dissipates the ability of the crack tip to propagate. The result is an increase in the K_{1C} fracture toughness of the grain of an order of magnitude.

The technical superiority of a fused grain of zirconia over one of alumina especially at very coarse sizes for rough grinding was recognized by the mid-1960s but was cost prohibitive, although alumina–zirconia blends showed advantage [39]. It was also recognized that an alumina–zirconia eutectic produced a strong structure due to a uniform dispersion of fine zirconia crystals in an alumina matrix. However, any excess alumina or zirconia from the eutectic would grow to a considerable crystal size depending on cooling rates from the melt. Rapid quench was therefore identified as a necessary pre-requisite of a processing route [40]. This required quench rates of 100°C/s, two orders of magnitude faster than the previous one. Numerous attempts were made during the late 1960s and 1970s to develop a viable process [41–49], using various inert cooling media or hearth plates, but it was a process developed by Scott [50, 51] of the Norton company that was to prove commercially and technically effective. The process from the original patent is illustrated in Fig. 1.29.

Molten alumina–zirconia from an electric arc tilt furnace is poured into the relatively thin spaces between a plurality of relatively thick heat sink plates of graphite or iron as they pass underneath, before being emptied at a discharge station by the plates separating. The result is a very fine structure of α-alumina with high tetragonal zirconia content. The zirconia is in the form of rods (or platelets) which, on the average, are less than 0.3 μm in diameter. The solidified melt is made up of cells or colonies typically 40 μm or less across their width. Groups of cells having identical orientation of microstructure form grains which typically include from 2 to 100 or more cells or colonies [52]. Figure 1.30 shows a TEM micrograph that illustrates the fine, rod-like zirconia structures within the larger cell [53].

After solidification the material is comminuted by standard methods of crushing, milling, and sizing to product grain. The processing will lead to some conversion of

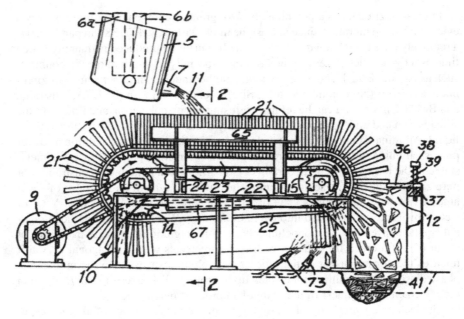

Fig. 1.29 Scott's patent detail for producing rapid quench alumina–zirconia

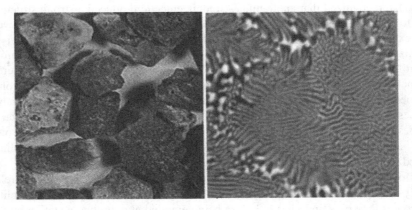

Fig. 1.30 Fused alumina–zirconia grain and a TEM micrograph of its rod-like zirconia structures [53]

the tetragonal to the monoclinic phase depending on the processing energy and especially on the final grain size. The smallest grain will lose much of its tetragonal toughening favoring this type of grain for use in coarse-sized, roughing operations.

The vast majority of alumina–zirconia grain for grinding wheels contains 25% zirconia, and sold under the trade names of ZF or ZS alundum, depending on the comminution method, and used in hot pressed resin bonds for rough steel, titanium,

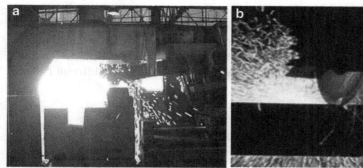

Rough grinding of Titanium billets   Cut off of foundry bar stock

Rail track grinding (Courtesy Saint-Gobain Abrasives)

Fig. 1.31 Examples of (dry) rough grinding using grinding wheels containing coarse alumina–zirconia grain

and nickel alloy billet conditioning or for foundry snagging. Grain size can be as coarse as 4# (0.26" or 6.8 mm) and used either as a single grain type or blended to include extruded sintered brown alumina (for finish), SiC (grinding titanium), or regular alumina. Billet conditioning is very aggressive form of grinding. The operation is run dry with wheel speeds up to 80 m/s (16,500 sfpm) and spindle power as great as 500 Hp on the most modern equipment. The workpiece is still often red hot from the furnace. Metal removal rates are extraordinary and can exceed 2,500 lbs/h ($Q \approx 40$ in.³/in./min or 400 mm³/mm/s) on steel or 400 lbs/h ($Q \approx 12$ in.³/in./min or 120 mm³/mm/s) on titanium, far exceeding most other metal removal processes. Other applications include the re-grinding of rail track in situ to remove fatigue cracks using special trains traveling at speeds of up to 6 mph (10 km/h). Grinding dry, in all these processes, the grains self-sharpen from cracks generated by thermal shock (Fig. 1.31). What is perhaps most striking about these processes is that the wheels are heavily reinforced with binder and grain and is processed at less than 200°C, yet the interface temperatures in grinding can far exceed 1,000°C [108] to the point of melting the metal. It is only the extremely poor thermal conductivity of the grain that prevents the rapid degradation of the bond around the grain base while the high temperatures erode the bond between the grains at the grinding interface to give chip clearance. Eutectic alumina–zirconia grain containing 40% zirconia is also produced and sold under the brand names of NZPlus, NZ®, NZP®, and Norzon®. It is used primarily for coated applications and as such requires a different balance of hardness (−) and toughness (+)

1.5   Engineered Alumina Abrasives

Engineered abrasive has microstructures that have been produced with controlled crystallite sizes from the submicron to micron level by processes other than simple fusion and communition. These include sol-gel/sintering and agglomeration techniques. The result is a family of grain types that micro-fracture at controlled micron or submicron levels and have the ability to be micro-trued, enhancing both the wheel life and process control compared with fused aluminum oxide grain.

1.5.1   "Ceramic" Sol-Gel Based Abrasives

The development and commercial success of first the sintered extruded alumina family of grains and then the rapid chilled fused alumina–zirconia grain had a major impact on the research programs of abrasive manufacturer in regard to the importance of control of grain crystal size. Furthermore, for alumina grain it was known that reducing the crystal size from the macroscale equivalent to a single crystal per abrasive grain, common in fused material, to micron or ideally <0.5 μm crystalline structures significantly enhanced grain properties such as hardness (Fig. 1.32) [54].

The response was, rather than using traditional fusing or sintering processes with their general limitations on cooling and crystallization rates, to consolidate microstructures from finer building blocks by sintering well-dispersed submicron precursors by the so-called sol-gel route. This allowed the consolidation of an α-alumina-based submicron, highly homogeneous and fully densified grain structure. The starting point of this new process is the manufacture of Boehmite, γ-aluminum oxide hydroxide γ-AlO (OH), from a modified version of the Ziegler process originally developed for the production of linear alcohols [55]. The material is produced as a submicron, narrowly sized powder which when mixed with water and a suitable acid dispersant forms an agglomerate-free sol-gel of aluminum hydrate ($Al_2O_3 \cdot H_2O$) with a dispersant size of about 100 nm. The sol-gel is then dehydrated/shaped and sintered (Fig. 1.33).

The biggest hurdle to overcome in the sintering process was to maintain a uniform submicron crystal size and full densification. Firing of a sol-gel from a standard commercial Boehmite at 1,400–1,500°C produces a large amount of porosity and relatively large grains of up >1 μm in size. This is believed due to high activation energy to convert from a transitional τ to α-alumina phase resulting in infrequent

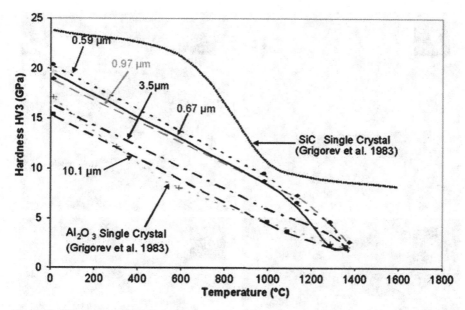

Fig. 1.32 Effect of crystal size on alumina grain hardness

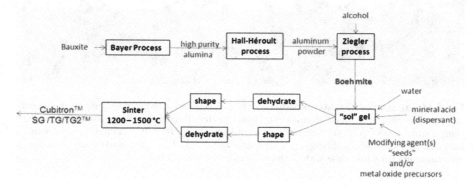

Fig. 1.33 Manufacturing route for the production of "Ceramic" alumina grain

nucleation with rapid uncontrollable growth rates. Attempting to control growth rates with lower temperatures, e.g., 1,200°C merely leads to larger crystals with higher porosity. There are two routes that have been developed to reduce the activation energy and control crystal size and densification. The first is the creation of a bi- or multi-composite structure through the use of modifying agents, the second is the controlled creation of a single α-alumina structure through the use of seeding agents (Fig. 1.34).

The earlier patents report the use of magnesia [56] which upon sintering forms a bi-composite structure of α-alumina plus a spinel structure of magnesium aluminate

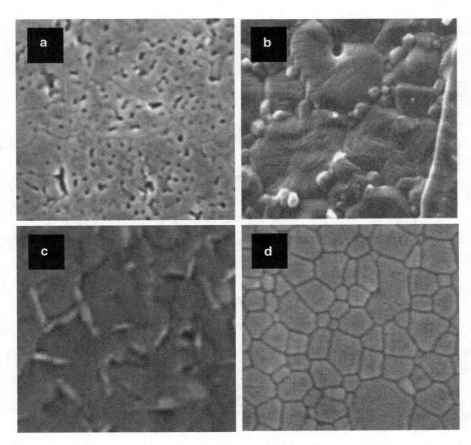

Fig. 1.34 (**a**) Sintered alumina microstructure from Boehmite with no modifying agent (image size 3 μm×3 μm); (**b**) Sintered alumina microstructure from Boehmite, magnesia modifying agent (image size 3 μm×3 μm); (**c**) Sintered alumina microstructure from Boehmite, magnesia, yttria, lanthana, and neodymia modifying agents (image size 1.5 μm×1.5 μm); and (**d**) Sintered alumina microstructure from Boehmite with seeding agent (image size 1.5 μm×1.5 μm)

at about 25% by volume as shown in Fig. 1.34b. Note the fine needle like spinel structure and the still relatively coarse α-alumina phase. This particular grain was used primarily for low force coated abrasive applications. Later, numerous patents report various multiphase systems using various modifying agents including zirconia, manganese oxide, chromia, nickel oxide, and numerous rare earth oxides. One particularly effective material contains magnesia together with yttria and other rare earth oxides such as lanthana and neodymia to produce a dense and hard (19 GPa) grain. In Fig. 1.34c, the microstructure shows a finer α-alumina phase (although still relatively coarse compared to the feed material) but with a submicron "magnetoplumbite" type structure of needles/plates formed from the modifiers [57].

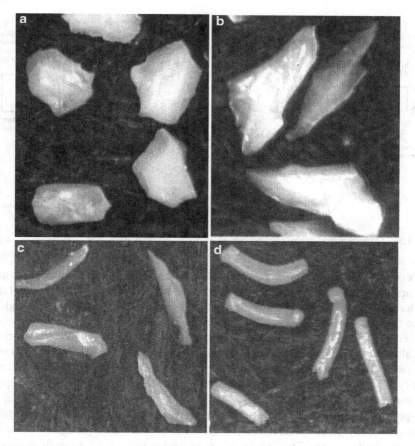

Fig. 1.35 (**a**) Tough blocky ceramic grain produced by milling; (**b**) Friable angular grain produced by crushing; (**c**) Weak extreme angular grain produced by crushing in green state; and (**d**) Extruded TG2™ ceramic grain (Courtesy Saint-Gobain Abrasives)

The structures created by the modifiers are believed to provide strength akin to rebar in reinforced concrete. This particular grain is the 3M™ Cubitron™ 321 grain [58].

The alternate route to controlling crystallization rates is by "seeding" the sol-gel with nano-sized (<100 nm) α-alumina or other materials with a crystallographic match to α-alumina such as α-ferric oxide or various titanates. Additions of 1–5% of seeding agent creates a heterogeneous nucleation condition by increasing the number of nucleation sites from $10^{11}/cm^3$ to $10^{14}/cm^3$, and an average crystal size of about 400 nm (Fig. 1.34d) [59, 60]. This type of grain is sold commercially under the name Norton SG™. One limitation of such fine crystal size is surface reactivity with standard vitrified bonds for fabricating grinding wheels. Bonds had to be developed to be fired at <1,000°C rather than the 1,200°C of older bonds used for fused alumina abrasives [61].

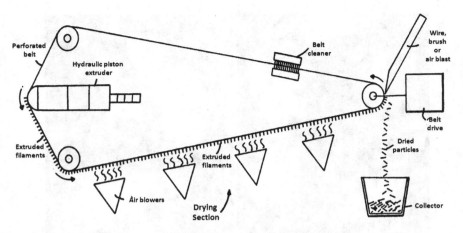

Fig. 1.36 Manufacturing process for producing extruded ceramic grain [62]

A comparison of Fig. 1.35a–d indicates that the single-phase seeded microstructure is finer than the multiphase microstructure and would be expected to be slightly harder and tougher. It would also be expected to give a longer life but require higher force to micro-fracture when used as abrasive grain, or should be used at a lower concentration in a blend. The multiphase grain would be slightly more free-cutting and also less reactive with high temperature vitrified bonds. These differences, however, are relatively minor compared to the difference in overall performance of this family of abrasives relative to fused alumina. Furthermore, further performance optimization can readily be obtained by wheel formulation and especially grain shape. Sol-gel manufacturing allows a much greater manipulation and control of grain shape. Standard crushing and milling methods can produce the typical strong blocky or weak angular shapes. The angularity can be further increased for careful processing of soft, dried pre-sintered material (Fig. 1.36). As expected these grains are also relatively weak but extremely successful if orientated on, for example, a coated application with relatively low grinding forces. More interesting, however, are novel techniques [62] that have been developed to extrude as rectangular prisms with extraordinary aspect ratios and having the appearance of smooth, surface defect free "worms" (Fig. 1.34d). Norton use TG™ grain with an aspect ratio of 5, and TG2™ with an aspect ratio of 8 [63]. Not only do these grains maintain a high toughness but they also have a very low packing density. Typical blocky grain may pack to about 50% by volume; an extruded grain with an aspect of 8 has a packing density closer to 30%. This provides for a very high level of permeability and excellent coolant access in the final fabricated wheel. Due to the toughness, shape, and ability to provide coolant the stock removal capabilities on tough superalloys such as Inconel or Rene alloys exceed that of cBN grain, for example, by an order of magnitude.

The most recent variant on the SG type abrasive is a grain called Quantum™ which maintains the submicron crystallite size and associated hardness of the SG abrasive family of grains but has controlled the levels of inclusions to promote

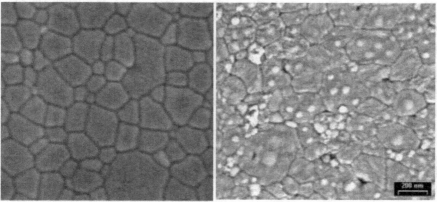

Norton SG Grain micro-structure   Norton NQ Grain micro-structure

Fig. 1.37 Comparison of Norton SG and Quantum grain microstructure

micro-fracture the lower force levels (Fig. 1.37). This also allows the grain to be micro-trued with dress depths in the 5–15 μm range to generate sharp, fractured, but durable cutting edges.

1.5.2 Agglomerated Grain

So far this discussion has covered grain produced by a fusion route followed by comminution and by chemical precipitation, sintering, and comminution. The former process creates grains with crystallites comparable in size to the grain, i.e., 50–200 μm, while the latter creates crystallite sizes in 0.2–5 μm range. The latest edition to the engineered grain family is to produce "agglomerated" grain made by a fusion, comminution, agglomeration, sintering, and recomminution. The resulting grain has a controlled crystallite size that bridges the gap between SG and fused. Since the size, shape, and chemistry of the crystallites are controlled by the initial comminution process, the possible variations in resulting grain grinding properties are enormous. Furthermore, the options for blending of SG, NQ, and Vortex grains in the same wheel offer an extraordinary range of complementary grain properties that are only just beginning to be optimized (Fig. 1.38).

As an example, it has been found that agglomerated grains pack to give a naturally high level of porosity in the resulting wheel structure making them very attractive for creep feed grinding. In addition the creation of very sharp crushed crystallites in the initial comminution process combined with controlled strength in the agglomeration binder allows controlled crystallite break-out to limit wear flat formation, resulting in a very cool grind on heat sensitive materials. This makes agglomerated grain based wheel structures a very promising replacement for SiC.

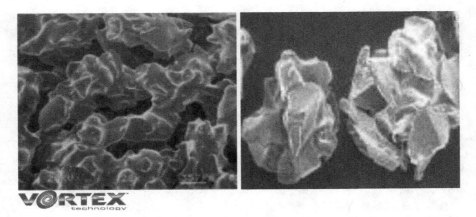

Fig. 1.38 Norton Vortex ™ agglomerated aluminum oxide grain

1.6   Superabrasives

Superabrasive refers to natural and synthetic diamond, and Cubic Boron Nitride (cBN) materials characterized by having the most extreme hardness and thermal conductivity. Their manufacture, or in the case of natural diamond their extraction from the ground, is also extremely expensive resulting in material cost greater than 1,000 times that of conventional grain . The processing and use of superabrasive grain is in some ways very different to conventional abrasive but in other very similar if more exacting.

1.6.1   Diamond

Natural diamond—Natural diamond has been a familiar part of industry since its foundation and still represents a surprising portion (>50 M carats/annum) of the machining business primarily for cutting tools, dressing tools for wheels and wear parts with secondary use as actual abrasive for grinding wheels or coated belts. Demand continues to be strong [64], but the top quality product for industrial use is often under economic pressure from the jewelry trade and low labor costs for polishing small gem quality stones in countries such as India leading to scarcity [65]. Nevertheless natural diamond, generally of a color, shape, or inclusion level unpopular for the gem business, remains the standard for single point dressing tools and stones in rotary dressing form rolls. Some crushed natural diamond is used in grinding wheels particularly in plated single products requiring grain with extreme sharpness and high angularity.

Natural diamond is formed at depths of 150–200 below the earth's surface under extreme temperature and pressure in the mantle. It may then be carried up in molten ⁻kimberlite and lamproite rock where it is found at the earth's surface within

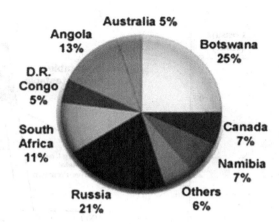

Fig. 1.39 Global supply of natural diamonds as of 2011

carrot-shaped "pipes," or alluvial deposits produced from erosion of these pipes most commonly within old landmasses known as cratons. Large diamonds of the size used in dressing tools are believed to form and remain over great period of time in the mantle but microdiamonds <0.5 mm are believed to form in the kimberlite and lamproitic magma [66]. Not surprisingly each individual pipe can produce a very unique range of diamond sizes and shapes. Some fields may contain predominantly microdiamonds that until recently were uneconomic due to the lack of traditional gem quality material. (Yields of diamond are of the order of <1 tonne per 13 million tonnes of ore processed.) Furthermore most of the major diamond deposits are in politically unstable areas of the world especially South and Central Africa although mines in NW Australia and Canada have recently come on line while Russia has produced large quantities of both gem and industrial diamonds for many decades (Fig. 1.39).

Synthetic HPHT diamond—The supply of natural diamond has for several decades lacked the consistency, security of supply, and cost position necessary to meet the requirements of modern industry with its increased demands to machine carbide, ceramics, and other advanced materials. Interest in synthesizing diamond was spurred after WW2 with the introduction of carbide tooling and the need for efficient means of fabricating them. The first to achieve synthesis of diamond was ASEA AB, Sweden, under the Research Directorship of Erik Lundblad in 1953. Independently in 1954 GE's "Super-Pressure Team" including Tracy Hall and Bob Wentorf produced its first synthetic diamond crystal which was the first to be repeatable and published. (See Finer points 2005 Superabrasive Resource Directory for further details [64, 67]).

Diamond is created by the application of extreme high temperatures and pressures to graphite. The stable form of carbon at room temperature and pressure is graphite with its familiar layered hexagonal lattice structure. Although bonding within the lattice is strong sp^3, covalent bonding between the layers is Van de Waals forces only, resulting in easy slippage and low friction. Diamond, which is metastable at room temperature and pressure, has a cubic arrangement of atoms with pure sp^3 covalent bonding with each carbon atom bonded to four others. The phase diagram for diamond/graphite is shown in Fig. 1.40.

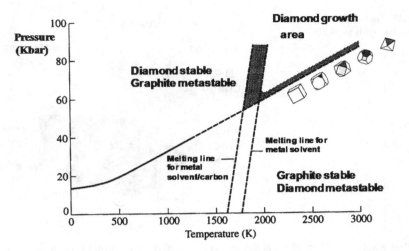

Fig. 1.40 Phase diagram for diamond and graphite

The direct conversion of graphite to diamond requires temperatures of 2,500 K and pressures of >100 kbar. Diamonds produced by this route are termed high pressure, high temperature (HPHT). The severity of the growth conditions can be reduced significantly by the use of a metal solvent such as nickel or cobalt. Graphite has a higher solubility in these solvents than diamond; therefore, at the high process temperatures and pressures the graphite dissolves in the molten solvent and diamond then precipitates out. The higher the temperatures, the faster is the precipitation rate and the greater the number of nucleation sites.

The earliest diamonds were grown fast at high temperatures and had weak, angular shapes with a mosaic structure. Also, the principal crystallographic planes of diamond are the cubic (100), dodecahedron (011), and octahedron (111). The relative rates of growth on these planes are governed by the temperature and pressure conditions and by the presence of metal solvent. In general at low temperatures the primary growth plane is cubic, while at the high temperatures it is octahedron. Careful control of the growth conditions allows the shape to be engineered to specific applications. In particular the blockiest, strongest form is the intermediate cubo-octahedral used in the strongest metal bonds for cutting or grinding concrete and glass (Fig. 1.41).

High temperature and pressures are generated by three main press designs: the belt press, the cubic anvil press, and the split-sphere (BARS) press. The belt press as developed for the first diamond synthesis by GE consists of an upper and lower anvil applying pressure to a cylindrical inner cell or bombe. The pressure is confined radially by a steel belt. Belt presses with substantial bomb volumes have been developed in recent years for the growth of large single crystals by companies such as Sumitomo and Ladd [68]. The bombe is doped with a seed crystal and a temperature gradient is created within such that diamond is gradually and steadily deposited over a prolonged period of time. The resulting diamond crystal is then laser cut

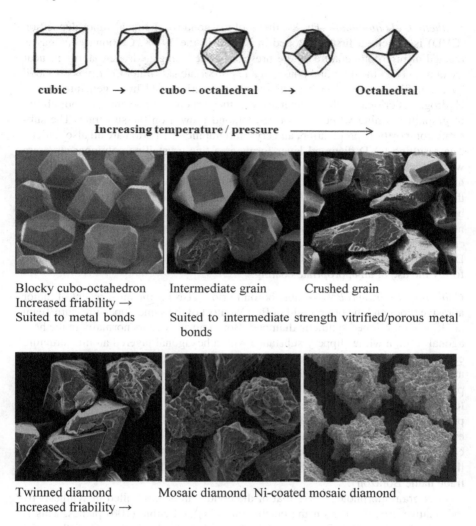

Fig. 1.41 Examples of synthetic diamond shapes and morphologies

along specific crystallographic directions to produce needles and blocks suitable for diamond dressing tools, rolls, and wire drawing dies.

The cubic anvil press has six anvils that apply pressure simultaneously onto the faces of a cube-shaped bombe. This type of press has a relatively small bombe volume and is best suited to fast processing time of medium to high friability diamond. The labor input required is relatively high but is popular for the recent rapid increase in production of diamond in China. The split-sphere BARS apparatus, developed in Russia, is a method for growing primarily large high quality diamond for specialized applications and the gem market.

Synthetic CVD diamond—The synthesis of diamond by chemical vapor deposition (CVD) is a method first developed in Russia in the 1970s. Carbonaceous gas is reacted at high temperature in the presence of reducing hydrogen atoms in near vacuum to form the diamond phase on an appropriate substrate. Energy is provided by a hot filament or plasma to dissociate the carbon and hydrogen into atoms. Hydrogen is critical in that it interact with the carbon and prevents any possibility of graphite forming while promoting diamond growth on the substrate. The substrate composition, preparation, and crystallographic orientation are all also critical. The resulting CVD diamond layer forms as a fine crystalline columnar structure with a thickness of up to 1–3 mm. There is only limited crystallographic orientation, making wear characteristics much more uniform and less sensitive to orientation than single crystal diamond. CVD diamond is not used as an abrasive but is again is becoming very prevalent in dressing tools and form rolls. CVD diamond contains no metal solvent contaminates which can actually be a problem when being fabricated for applications such as shaped cutting tools, since it cannot be EDM wire cut. Diamond wetting also appears more difficult in brazing and must be compensated for by the use of an appropriate coating.

Cubic boron nitride (cBN)—Cubic boron nitride (cBN) is the most recent of major abrasive types and the brain-child of Bob Wentorf of the same Super-Pressure Team at GE who developed synthetic diamond. Boron nitride occurs normally in the hexagonal form, a white slippery substance with a hexagonal layered atomic structure called h-BN (or α-BN) similar to graphite but with alternating nitrogen and boron atoms. Wentorf noted its similarities to the structure and bonding of graphite and proceeded to determine a suitable high temperature solvent to grow the cubic structured form—cBN (or β-BN). cBN is not found naturally but must be synthesized at pressures and temperatures comparable to those for diamond. The chemistry, however, is quite different; cBN shows no affinity for transition metals. Instead the successful solvent/catalysts are metal nitrides, borides, and oxide compounds of which the commonest is Li_3N. cBN was introduced commercially by GE in 1969 under the trade name Borazon®.

cBN grain morphology, like that of diamond, can be controlled in synthesis by the relative growth rates on the octahedral (111) and cubic (100) planes. This is controlled again with temperature and pressure but also be doping. Growth on the (111) planes dominate but because of the presence of both boron and nitrogen in the lattice, some (111) planes are terminated by boron atoms and some by N atoms. In general boron (111) plane growth dominates and the resulting crystal morphology is a truncated tetrahedron but twinned plates and octahedra are also common (Fig. 1.42). In addition shape can also be driven towards the octahedral or cubo-octahedral morphologies. The net result is there is a much wider potential availability of grain shape to choose from than for diamond in addition to twinned and multicrystalline material. Pure, stochiometrically balanced boron nitride in the cubic form is colorless but commercial abrasives are various shades of amber color through brown to black depending on the level and type of dopants present. The black color in particular is believed to be due to an excess of boron (Fig. 1.43).

Fig. 1.42 cBN crystal growth morphologies [69]

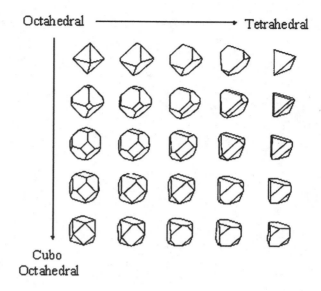

Octahedral ⟶ Tetrahedral

Cubo Octahedral

Fig. 1.43 cBN commercial grain examples [69]

Chapter 2
Vitrified Bonding Systems and Heat Treatment

2.1 Vitrified Bonding Systems

Grinding wheels of various types, sizes, and shapes are geometrically defined bodies consisting of abrasive grains bonded with various types of ceramic bond. During heat treatment, or firing, the alumina and bonding minerals react with the bond, and this has an important effect on the properties of the resultant tool. For this reason, studies of the structure of abrasive tools made of alumina, and the determination of the phase composition and structure of their bonds are important research tasks. Generally speaking, grinding wheels are made in the following way. The bond materials are ground until they can pass through a 120 mesh screen without leaving a residue, weighed, and mixed. The firing resistance of the bond is checked and it is then passed to the press shop. Here, alumina of the required size and the bond materials are weighed in the quantities specified by the formulation, moistened with a dextrin or sodium silicate solution, and thoroughly mixed in a planetary mixer. The finished molding material is carefully weighed into exact portions that are then poured into moulds and compressed to the required volume. The molded product is then dried at 90–100°C and fired. Heat treatment, or firing, takes place at temperatures of 1,250–1,300°C, in continuous or batch furnaces. The relatively long firing times (up to 80–120 h) are due firstly to the nonuniformity of abrasive tools, composed of abrasive grain and ceramic bond, and secondly to the need for the bond to become sufficiently mobile to coat the abrasive grains without disturbing the structure of the tool or deforming it. Firing time and, more importantly, the length of time spent in the preheating and cooling zones of the kiln also depend on the dimensions of the abrasive tools, and increase with their size. Fired grinding wheels are machined on lathes to the required shape and size. Finished abrasive tools undergo hardness tests, and grinding wheels are additionally rupture tested at 1.5–2 times their peripheral velocity.

M.J. Jackson and M.P. Hitchiner, *High Performance Grinding and Advanced Cutting Tools*, SpringerBriefs in Applied Sciences and Technology, DOI 10.1007/978-1-4614-3116-9_2, © Springer Science+Business Media New York 2013

2.2 Grinding Wheel Structure Formation During Heat Treatment

Ceramic bond materials are made of refractory and fusible clays, feldspar, quartz, talc, and sodium silicate. The chemical composition of ceramic bonds is typically 65–70% SiO_2, 20–25% Al_2O_3, and approximately 10% of alkali and alkaline earth metal oxides. Figures 2.1 and 2.2 give some idea of the structure of an abrasive tool. They are photographs of polished sections prepared from regular alumina grinding wheels (grit 46, K-hardness and medium structure (6)). The general appearance of a section is shown in Fig. 2.1. The micrograph demonstrates the nonuniform structure of the grinding wheel. In places, the alumina grains are surrounded with a light film of the bond material, with fine bridges linking the grains and large open pores, while elsewhere the grains seem to be immersed in the bond, which contains fine closed pores.

Detailed examination of individual areas of the section clearly demonstrates that the nonuniformity is much greater than can be seen at low magnifications. Most of the section consists of grains or fragments of monocrystalline alumina (corundum). The grains are usually surrounded by fine rims of the bond material, which form bridges between the corundum grains. Adjoining bridges are separated by large round pores with smooth internal surfaces (Fig. 2.3). Aggregate grains, consisting of several corundum crystals cemented by slag interlayers, are broken down by the bond during firing, and individual corundum crystals become immersed in the glass, which frequently contains crystals produced by the interaction between the accessory minerals of alumina and the bond (Fig. 2.4).

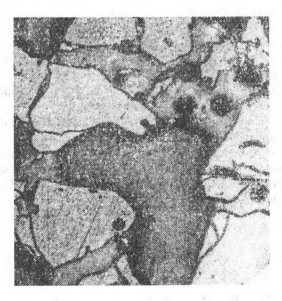

Fig. 2.1 Abrasive tool body: dense alumina aggregates cemented with a vitreous bond. Reflected light, ×60 magnification (Courtesy of Professor Givi Bockuchava)

Fig. 2.2 A meniscus of vitreous bond material between grains. Reflected light, ×250 magnification (Courtesy of Professor Givi Bockuchava)

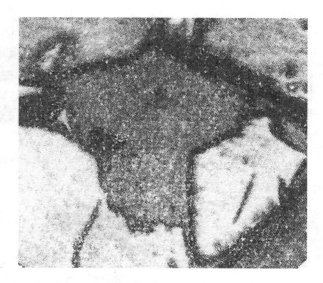

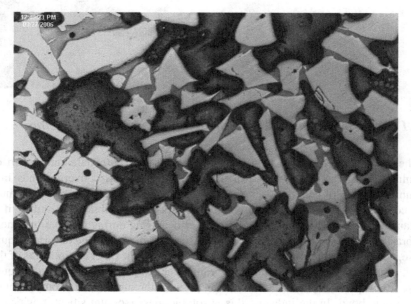

Fig. 2.3 A section of an abrasive tool showing adjoining bridges are separated by large round pores with smooth internal surfaces: an aggregate alumina grain broken down by the bond. The bond contains rutile aggregates (*white*), pores and microcracks (*black*). Reflected light, ×250 magnification

Therefore, abrasive tools appear to be the most complex of silicate products, and it is quite evident that their properties, like those of any other silicate product, will be determined by their phase composition, the structure of the bond, the strength of its adhesion to the alumina grains.

Fig. 2.4 A section of an abrasive tool: an aggregate alumina grain broken down by the bond. The bond contains rutile aggregates (*white*), pores and microcracks (*black*). Reflected light, ×250 magnification (Courtesy of Professor Givi Bockuchava)

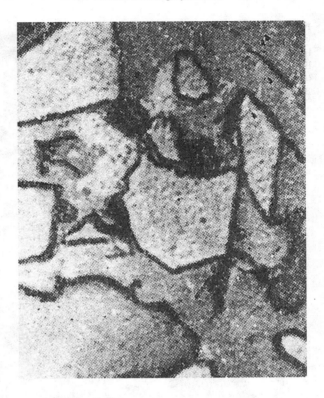

2.2.1 Physicochemical Processes That Occur During Firing

Studies of ceramic bonds used in industry, both in their pure state and as part of alumina abrasive tools, have shown that the firing process converts bonds into nonuniform glass of complex composition, which on devitrification produces mullite, cordierite, spinel if the bond contains MgO, and anorthite and mullite if it contains CaO. The physicochemical processes occurring during the heat treatment of abrasive tools can be described as follows. Firing causes the bond to form a liquid phase, and as the temperature rises, this phase dissolves the other components of the bond, as well as the surface of alumina grains. The temperature at which the liquid phase first appears depends on the melting point of the most fusible component of the ceramic bond. In the industry, this is soluble glass (sodium silicate), which is either introduced into the bond in its solid state, or forms part of the solution used to wet the abrasive material during molding. Undoubtedly, if other oxides were also introduced into the bond in the form of more fusible compounds, they too would act as solvents and would react with the alumina.

However, because alumina is relatively soluble in alkali oxide components, the most active solvent is Na_2O. Since sodium silicate becomes quite mobile even at 650°C, it can be assumed that the dissolution of alumina (and of bond components) begins at this temperature. Initially, the concentration of Na_2O in the newly formed

Table 2.1 Fired chemical composition of several bonds and their firing resistance

Bond no	Chemical composition, wt%									Firing resistance (°C)
	SiO_2	TiO_2	Al_2O_3	Fe_2O_3	CaO	MgO	K_2O	Na_2O	Total	
1	67.68	1.34	16.58	2.40	1.45	0.54	Nil	9.91	99.88	1,150
2	67.73	0.80	12.50	1.70	1.40	0.60	–	14.80	99.53	970
3	67.48	Nil	3.04	1.27	1.40	0.82	–	25.74	99.75	650
4	69.14	1.14	11.31	1.91	5.40	0.62	–	10.14	99.66	1,060
5	67.92	1.20	11.56	1.94	1.25	5.10	–	10.66	99.63	1,090
6	67.72	0.45	13.02	2.36	0.40	0.56	5.43	10.08	100.02	1,080

Table 2.2 Determination of Al_2O_3 in these bonds, extracted from the samples after heat treatment at various temperatures

Bond no.	Al_2O_3 content (%) of bonds heat treated at various firing temperatures (°C)						
	450	650	850	1,050	1,250	1,250[a]	1,250[b]
1	16.62	17.76	20.25	23.18	22.45	25.47	26.50
2	12.58	16.73	21.08	23.16	24.28	26.13	28.07
3	3.06	4.67	10.52	23.57	25.98	28.34	28.98
4	11.34	16.25	19.53	22.73	25.52	26.21	26.48
5	11.57	13.44	18.94	23.53	26.08	29.98	30.61
6	13.12	15.58	18.76	23.08	24.33	26.46	27.44

[a]Holding at the final firing temperature (8 h)
[b]Holding at the final firing temperature (16 h)

liquid phase is relatively high, up to 25%. As the other bond components and alumina dissolve (as the temperature increases), the concentration of Na_2O in the melt falls, its viscosity rises, and at the holding temperature dissolution is virtually complete.

The interaction between alumina and the bond is illustrated by the figures contained in Tables 2.1 and 2.2. Table 2.1 gives the fired chemical composition of several bonds and their heat resistance, or temperature at which melting first occurs (also known as the bond softening temperature). Table 2.2 contains the results of determination of Al_2O_3 in these bonds, extracted from the samples after heat treatment at various temperatures. The tables demonstrate that there is a direct relationship between the amount of dissolved alumina, the final firing temperature, and the alkali content of the bond.

Irrespective of its initial value, the heat resistance reached by the bond material during the firing process is close to the abrasive tool holding temperature. This explains, why abrasive products can be fired at temperatures 200°C and even 300°C higher than the bond softening temperatures, without the bond leaking and the products being deformed. It is interesting to note that bond 5, which contains magnesium oxide and has high initial heat resistance, dissolved alumina faster than all the others. This is explained by the formation of spinel at the contact interface with alumina, which prevents rapid increase of the alumina content of the liquid phase and thus allows it to retain greater mobility.

As the material dissolves, new mullite, anorthite, cordierite, and spinel crystals appear in the liquid phase at 1,200°C. Thus, the first heating stage of the firing process is the phase of chemical reactions and physicochemical transformations through which the bond reaches a state of equilibrium and partial crystallization. During the holding period, the composition of the bond equalizes and new crystals appear. This is conditional on the local dissolution of alumina during the heating period causing its composition to move into the area of molten bond that is capable of devitrification. The holding period is therefore chiefly a period when new compounds formed in the bond crystallize out of the liquid phase.

During the cooling process, the viscosity of the bond increases so much that the crystallization of new compounds virtually ends. This is why the cooling pattern seems to play an important part, not in the sense of giving the bond a specific structure, but in the sense of preventing mechanical stresses arising in grinding wheels due to their heterogeneous composition, and, more importantly, due to the simultaneous presence of a crystalline component, i.e., alumina grains and the vitreous bond. The cooling period is essentially a period when physical changes take place in the grinding wheel. Studies have shown that the vitreous content of the bond determines the mechanical properties of the tool. The vitreous content of the bond increases from 75 to 100%, with resulting breaking strength increases from 100 to 200 kg/cm².

Of the minerals studied, the only one that appears to improve the mechanical properties of grinding wheels is spinel $MgO \cdot Al_2O_3$, which is formed at the contact of the bond with alumina and encloses its grains in a casing of fine octahedra, no larger than 8 μm in diameter. The Na_2O contained in the tool dissolves the alumina grain, forming a small area of melt enriched with Al_2O_3, thus aiding the formation of spinel. The dissolution of up to 4% of alumina in the bond increases the mechanical strength of the grinding wheel, provided that the bond retains its vitreous structure, or small quantities of minerals form at the contact interface.

Ceramic bond materials producing abrasive tools with good mechanical properties are those located near the SiO_2 apices of the tetrahedra of two systems: Na_2O–K_2O–Al_2O_3–SiO_2 and Na_2O–MgO–Al_2O_3–SiO_2, containing these materials in the following proportions: $SiO_2 = 70$–75%; $(K_2O)MgO = 5\%$; $Al_2O_3 = 15$–10%, and $Na_2O = 10\%$. In modern grinding wheel firing conditions, these compounds react vigorously with alumina. In this process, the bond is enriched with Al_2O_3, whose content (in the four-part system) rises to 30–35%. These compounds form glass that does not devitrify during firing and produces grinding wheels with breaking strengths of 170–200 kg/cm².

2.2.2 Ceramic Bond Minerals That Form During Firing

Alongside anorthite, mullite, cordierite ($2MgO \cdot 2Al_2O_3$–$5SiO_2$), and spinel, which form in the bond when it is enriched with alumina, firing gives rise to other new formations, produced by the interaction between the bond and the accessory minerals

Fig. 2.5 A bond interlayer separating alumina grains and consisting of plagioclase (*gray*) with anatase interpenetrations (*white*). Reflected light, ×300 magnification (Courtesy of Professor Givi Bockuchava)

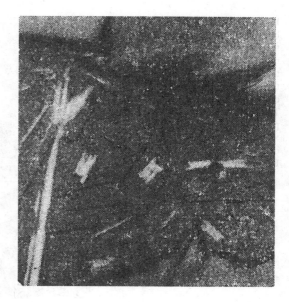

Fig. 2.6 Mullite (*pale gray*), anatase (*white elongated sections*), magnetite (*white dendrites*), and rutile (*round white formations*) in glass separating fractured alumina grains. Reflected light, ×250 magnification (Courtesy of Professor Givi Bockuchava)

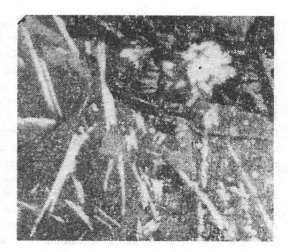

present in alumina. They include plagioclases, anatase, hematite, magnetite, and rutile. Let us now consider the process of formation of each of these minerals.

Anorthite glass (slag) contained in regular alumina grains in the form of streaks or interlayers is dissolved by the bond. The composition of the bond is significantly changed as a result, and on crystallization it forms plagioclase and anatase (Fig. 2.5). Silica-rich slag is also absorbed by the bond, but does not disturb its vitreous structure. The resultant sections of the grinding wheel usually consist of glass containing small quantities of mullite prisms, acicular rutile crystals, and magnetite dendrites (Fig. 2.6).

Fig. 2.7 A portion of the bond between alumina grains: the formation of granular rutile aggregates (*white*) through the oxidation of titanium carbide (*white*) embedded in the alumina. The bond contains large pores (*white*). Reflected light, ×250 magnification (Courtesy of Professor Givi Bockuchava)

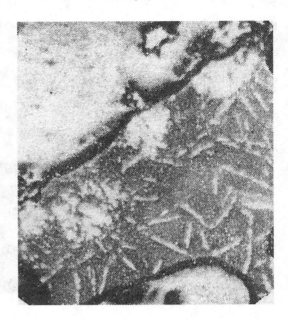

Titanium oxides occurring in glass present as accessory minerals in alumina convert into disperse rutile grains on firing. The grains are then recrystallized in the bond, forming acicular aggregates. Anosovite behaves differently, converting into a pseudo-morph after anatase and appearing in the bond in the form of brown colored crystals. Titanium carbide and nitride oxidize during firing forming granular rutile aggregates. Polished sections distinctly show the explosive nature of their oxidation and its adverse effects—presence of crystallization of the rutile in the bond and the formation of gas bubbles (Fig. 2.7). In addition, during the firing process, the action of Na_2O present in the bond causes the solid solution of Ti_2O_3 in alumina to break down and a fringe of rutile needles form on the surfaces of the alumina crystals (Fig. 2.8).

Measurements have shown that the solid solution of Ti_2O_3 in alumina crystals break down to a depth of 30–40 μm. The presence of ferroalloy has a dramatic effect on bond composition. The bond material surrounding the ferroalloy bead is saturated with hematite and magnetite formed through ferroalloy oxidation (Fig. 2.9). Titanium sulfide and carbide present in industrial monocrystalline alumina grains also convert to rutile during firing, forming granular aggregates distributed in the bond. A grinding wheel bond usually contains up to 20–25% of neocrystalline formations, with crystal sizes not exceeding 40–50 μm. If the bond is overfired, these crystals reach 80–120 μm and develop microcracks in the bond (Fig. 2.10). The features of minerals formed in grinding wheels after firing, which form the basis of microscopic analysis, are described below.

Anorthite ($CaO \cdot Al_2O_3 \cdot 2SiO_2$) appears in the bond in the form of randomly arranged colorless lamellae and columnar (elongated tabular) crystals displaying negative elongation and polysynthetic twinning. Moderate refractive index and birefringence: $N_g = 1.588 \pm 0.002$, $N_p = 1.575 \pm 0.003$, $N_g - N_p = 0.013$.

Fig. 2.8 A fringe of rutile needles at the alumina-bond contact. Reflected light, ×250 magnification (Courtesy of Professor Givi Bockuchava)

Fig. 2.9 Ferroalloy and iron oxides in bond material. Reflected light, ×250 magnification (Courtesy of Professor Givi Bockuchava)

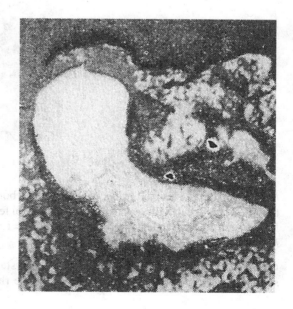

Fig. 2.10 An anorthite spherolite in bond material showing microcrack formation. Analyzer out, ×200 magnification (Courtesy of Professor Givi Bockuchava)

Fig. 2.11 Mullite in bond material. Analyzer out, ×80 magnification (Courtesy of Professor Givi Bockuchava)

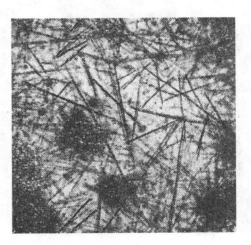

Mullite ($3Al_2O_3 \cdot 2SiO_2$) is always colorless in the bond (Fig. 2.11). It crystallizes in the form of fine needles, mostly gathered into felted and radiated aggregates. Moderate refractive index and birefringence: $N_g = 1.654 \pm 0.002$, $N_p = 1.642 \pm 0.003$, $N_g - N_p = 0.012$.

Cordierite ($2MgO \cdot 2Al_2O_3 \cdot 5SiO_2$ (Fig. 2.12)) crystallizes in the form of colorless short prismatic pseudohexagonal crystals (in the rhombic system), which have a

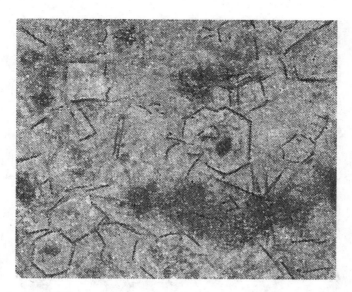

Fig. 2.12 Cordierite in bond material. Analyzer out, ×160 magnification (Courtesy of Professor Givi Bockuchava)

moderate refractive index and low birefringence: $N_g = 1.525 \pm 0.003$, $N_p = 1.521 \pm 0.002$, $N_g - N_p = 0.004$. In polished sections, it is dark gray, with low reflectivity and a relief similar to that of glass.

Spinel ($MgO \cdot Al_2O_3$) is yellow or colorless in transmitted light (Fig. 2.13). It appears in the form of octahedra and grains at the alumina contact. It has a high refractive index: $N = 1.722 \pm 0.003$.

Plagioclases crystallize in the triclinic system. In transmitted light, they appear in the form of elongated tabular crystals, frequently displaying polysynthetic twinning. Moderate refractive index and birefringence: $N_g = 1.553 - 1.558$, $N_p = 1.547 - 1.552$. In polished sections they are gray, with low reflectivity and relief equal to that of as glass.

Hematite (Fe_2O_3) crystallizes in the trigonal system. In transmitted light, it appears in the form of orange-red irregular accumulations, less often in the form of hexagonal or triangular lamellae. High refractive index and birefringence: $N_g = 3.01$ Li, $N_p = 2.78$ Li. The irregular accumulations result from ferroalloy oxidation, while the regular lamellae, frequently with regular orientation, appear on the surfaces of alumina crystals (Fig. 2.14). They are formed by the recrystallization of ferrous oxide film on alumina grains (in monocrystalline alumina). In reflected light, hematite is white, with above-average reflectivity and a higher relief than glass.

Magnetite (Fe_2O_4) crystallizes in the cubic system. It is opaque. In transmitted light it appears in the form of irregular black accumulations associated with ferroalloy, less often in the form of fine cubic crystals and skeletal cruciform dendrites. In reflected light, it is white with high reflectivity and a higher relief than glass.

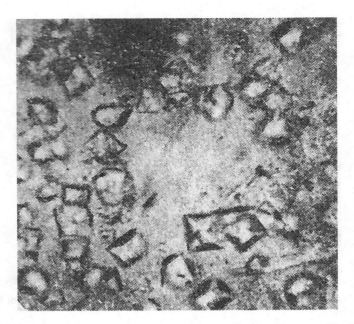

Fig. 2.13 Spinel in bond material. Analyzer out, ×250 magnification (Courtesy of Professor Givi Bockuchava)

Fig. 2.14 Hematite on the alumina-bond contact. Reflected light, ×150 magnification (Courtesy of Professor Givi Bockuchava)

Rutile (TiO$_2$) crystallizes in the tetragonal system. It is colorless, and forms either granular formations or acicular or prismatic (frequently hollow) crystals. Its refractive index and birefringence are extremely high: $N_g = 2.903$, $N_p = 2.616$. In reflected light, it appears as either sinuous lacy aggregates and accretions formed by the oxidation of titanium carbide and nitride, or in the form of white rectangular and rhomboid sections, formed by the alteration of anosovite. It has high reflectivity and a relief higher than glass.

Anatase (TiO$_2$) crystallizes in the tetragonal system, forming prismatic and rod-like crystals, which can be either colorless or colored yellow or brown. Its refractive index and birefringence are very high, $N_g = 2.56 \pm 0.02$, $N_p = 2.48 \pm 0.02$. In reflected light, it appears in the form of white, rectangular, and rhomboid sections. It has above-average reflectivity and a relief higher than glass.

The following case studies highlight the importance of interface compounds and bonding phases on grinding wheel wear in high-performance vitrified grinding wheels. It should be noted that these studies are focused on clay-based bonding systems.

2.3   Case Study I: Interfacial Compounds and Their Effect on Grinding Wheel Wear

The type of grinding wheel considered in this case study is made using aluminum oxide (α-Al$_2$O$_3$), a hard material with a Knoop hardness of up to 2,000 kg/mm^2, is used in the grinding industry in two principal forms: a high purity, fused form of alumina containing over 99.9 wt% Al$_2$O$_3$ that is white in appearance; and a fused, brown colored, alumina of 95 wt% purity. The main impurity in this latter form is TiO$_2$ at a level no greater than 3 wt%. This tends to increase the toughness of the grain and is accompanied by other impurities such as MgO, CaO, Fe$_2$O$_3$, and ZrO$_2$. Other grinding wheels described in this case study use cubic boron nitride (CBN) that has a Knoop hardness in excess of 4,500 kg/mm^2.

The range of vitreous bonding systems and abrasive types is very large, though only alumino-alkalisilicate and alumino-borosilicate bonding systems are used by the abrasive wheel industry. The normal practice is to adjust the proportions of Al$_2$O$_3$, B$_2$O$_3$, SiO$_2$, and alkali oxides to achieve the desired fluidity. Other chemical and physical properties can be modified by the addition of alkaline earth oxides. Vitreous bonds are composed of mixtures of quartz, feldspar, clay, borate minerals, and ground frits. In practice, the bonds are mixed with a variety of abrasive grains. However, this case study considers high purity and titanium-doped varieties (using a typical mesh size of 220, which is approximately 62 μm diameter abrasive grain size) and CBN with B64 grain size (approximately 63 μm in diameter).

The grinding process is accompanied by wear of the abrasive wheel, and the rate of this wear plays an important role in determining the efficiency of the grinding process and the quality of the workpiece. The structure of a vitrified grinding wheel

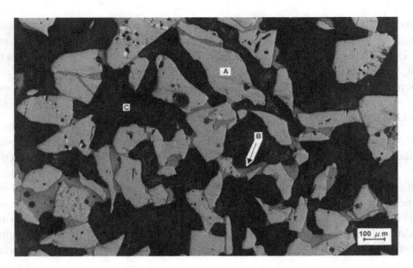

Fig. 2.15 Microstructure of a vitrified grinding wheel. A—denotes abrasive grain, B—denotes vitrified bonding phase, and C—represents distributed porosity

is composed of abrasive grains, a bonding system, and a large number of pores. Figure 2.15 shows a typical porous grinding wheel structure [70]. Krabacher [71] pointed out that wear mechanisms in grinding wheels appear to be similar to that of single-point cutting tools, the only difference being the size of swarf particles generated. The wear behavior observed is similar to that found in other wear processes; high initial wear followed by steady-state wear. A third accelerating wear regime usually indicates catastrophic wear of the grinding wheel, which usually means that the wheel will need to be dressed. This type of wear is usually accompanied by thermal damage to the surface of the ground workpiece. The performance index used to characterize wheel wear resistance is the grinding ratio, or G-ratio, and is expressed as the ratio of the change in volume of the workpiece ground to the change in the volume of the grinding wheel removed, thus,

$$G = \frac{\Delta v_w}{\Delta v_s} \tag{2.1}$$

Grinding ratios cover a wide range of values ranging from less than 1 mm³/mm³ for vanadium-rich high-speed steels to over 60,000 mm³/mm³ when internally grinding bearing races using CBN abrasive wheels. Attempts have been made on how to address the problems related to the wear of abrasive grains in terms of the theory of brittle fracture. The conclusions of various researchers lead us to believe that the variety of different and interacting wear mechanisms involved, namely, plastic flow of abrasive, crumbling of the abrasive, chemical wear, etc., makes grinding wheel

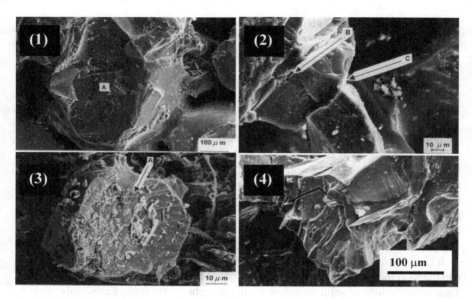

Fig. 2.16 Grinding wheel wear mechanisms: (1) abrasive wear—A denotes a wear flat generated by abrasion; (2) bond bridge fracture—A denotes the abrasive grain, B denotes the interfacial bond layer, and C denotes a crack passing through the bond bridge; (3) abrasive grain fracture—A denotes crystallographic grain fracture; and (4) interface fracture between abrasive grain and bond bridge

wear too complicated to be explained using a single theoretical model. High efficiency precision grinding processes place extreme loads onto the grain and the vitrified bonding bridges.

2.3.1   Wear Mechanisms

Four distinct wheel wear mechanisms that contribute to the wear of grinding wheels are identified as (Fig. 2.16):

1. Abrasive wear (formation of wear flats on the surface of abrasive grains)
2. Fracture of bond bridges
3. Fracture of abrasive grains due to mechanical and thermal shock loads
4. Fracture at the interface between abrasive grain and bond bridge

2.3.1.1   Abrasive Wear

The formation of wear flats on abrasive grains leads to a loss of grain sharpness. The sources of minute scale wear are:

1. Wear due to frictional interaction between workpiece and abrasive grain
2. Plastic flow of the abrasive grain at high temperature and pressure
3. Crumbling of the abrasive grain due to thermal diffusion and microscale mechanical impact
4. Chemical reaction between abrasive and workpiece material at elevated temperatures and in the presence of grinding fluids

The latter mechanism can reduce the resistance of the abrasive grain to other wear mechanisms. Dull abrasive grains are caused by the generation of wear flats on active grains that leads to an increase in the area of contact and frictional interactions between abrasive grain and the workpiece. At the point of dulling of the abrasive grain, very high temperatures exist in the area of contact that greatly enhances adhesion and chemical reaction between two surfaces. If grain and bond bridge fracture does not occur during grinding then the plateau area on the grain widens and the rate of wear increases. If fracture is delayed further, as with hard grinding wheels, then the wheel becomes glazed and the workpiece is thermally damaged.

It has been shown experimentally [72] that chemical affinity between the abrasive and the workpiece material can be used as a guide for the selection of grinding wheels. Their observations of solid diffusion of silicon carbide into ferrous materials explain the catastrophic wear rates exhibited by these "wheel–workpiece" combinations. The most common method used for measuring wear flat area employs an optical or an electron microscope. Hahn [73] observed and analyzed the effect of the increasing wear flat area during the plunge grinding of various workpiece materials. Hahn concluded that grinding forces gradually increase during wear-flat formation up to a point where the grinding wheel will restore its sharpness due to abrasive grain fracture.

2.3.1.2 Fracture Wear

The occurrence of abrasive grain and bond fracture are considered simultaneously for the following reasons:

1. They are of the same nature, i.e., fracture of brittle materials and hence the theory of brittle fracture is applicable to both bonding bridge and abrasive grain. The applied thermal and mechanical loads, usually under cyclic conditions, cause initiation and further development of cracks that leads to fracture and the formation of new irregular surfaces.
2. They are related to dressing methods used and occur simultaneously. The initial and final stages of wheel life between dressings exhibit fracture wear that is a combination of abrasive grain and bonding bridge fracture.
3. The relative amounts of bond bridge and abrasive grain wear cannot always be calculated. An investigation into precision grinding employed a soft wheel that gave a high percentage of bond fracture, whereas a harder wheel gave partial abrasive grain fracture. Wear by attrition occurred in both cases.

However, the combination of grinding parameters such as equivalent chip thickness and the grindability of the workpiece material determines the effective wheel hardness, and so no single feature of the grinding process can be used to predict the fracture pattern of the wheel in advance. The difficulty when relating grinding wheel wear due to fracture to a particular grinding condition arises from the lack of knowledge about the loads applied to both abrasive grains and their bonding bridges and their response to these applied loads.

Tarasov [74] suggests that abrasive grain fracture occurs as a result of mechanical forces due to chip formation, or thermal shock, induced by instantaneously high temperatures. Hahn [73] proposed a thermal stress hypothesis to explain the fracture of abrasive grains. Plunge grinding experiments were conducted under fixed normal load conditions. Hahn asserted that as wear progresses measurements of torque indicated that the tangential force decreases. This led to the conclusion that abrasive grain fracture due to mechanical loading will not occur. Mechanical stresses wear also considered as an explanation for the different rates of wear of the grinding wheels used in the experiments.

Bhattacharyya et al. [75] observed abrasive grain loss due to fracture using an electron microscope. They concluded that they could not differentiate between Peklenik's crystal splintering, i.e., grit flaking due to thermal stress, and abrasive grain fragmentation. However, they did explain their results in terms of Hahn's thermal shock hypothesis. Hahn's experimental conditions suggested that attrition of the abrasive was expected to occur through abrasive wear. Wear measurements by Hahn [73] were based on the reduction in grinding wheel diameter, which Malkin and Cook [76] attributed to abrasive wear. Wear rates recorded were of the order of 50 μin./s. It was expected that abrasive wear rates were in the region of 5 μin./s. This rate was observed under light grinding conditions. Under heavy grinding conditions, the conditions of wear appeared to be more complex.

Malkin and Cook [76] collected wheel wear particles for each grade of grinding wheel when grinding using a fixed set of operating conditions. They analyzed their size distributions statistically and discovered that a soft-grade grinding wheel (G-grade) produces 85% of grinding debris associated with bonding bridge fracture, whilst a harder-grade grinding wheel (K-grade) produces 55% of grinding debris associated with fractures of bonding bridges. Abrasive wear accounted for 4% of the total wear in both cases.

The strongest evidence in support of the idea of fracture due to mechanical loading is that fracture occurs some distance away from the cutting edge [77]. It was concluded that the heat generated by cutting has no effect on abrasive grain fracture since the peak temperature of the abrasive grain occurs at the surface of the grain in contact with the workpiece where fracture is initiated on cooling according to the thermal stress hypothesis. The hypothesis does not take account of any difference in the coefficient of thermal expansion between abrasive grain and bond bridges, and also the effect of thermal shocks on the quenching action of grinding fluids on the abrasive grain leaving the cutting zone. The latter case was analyzed and it was reported that the thermal stress in an abrasive grain due to a pulsating heat source showed that the magnitude of the maximum tensile stress is not large enough to cause fracture of the grain. Malkin and Cook [76] adopted the mechanical loading approach. Malkin and Cook [76] derived an expression from first principles for the probability of bond fracture in terms of a bond stress factor.

Although bond and grain fracture are similar mechanisms, they have a different effect on the economics of the grinding process. The first mechanism results in a rapid loss of the grinding wheel, while the second mechanism, on a comparable scale with the un-cut chip thickness, generates sharp cutting edges and is known as the "self-dressing action." Both mechanical and thermal stresses appear to be responsible for fracture wear. The effect of heat at the abrasive grain and workpiece interface is responsible for locally affecting the mechanical properties of the abrasive grain. However, fragments of larger sizes of abrasive grain are more likely to occur through mechanical loading that governs bond fracture and the self-sharpening action. A method of alleviating the onset of bond fracture due to unusually large mechanical loads is to dissolve deleterious particles in the bonding system that weakens the structure of the bonding bridge.

In vitrified bonds, these particles are quartz particles that naturally occur in ceramic raw materials. These particles reduce the load-bearing strength of the bonding bridges during vitrification heat treatment. The study of the effect of the elastic modulus on the fracture behavior of vitrified abrasive grinding wheels was conducted by Decneut, Snoeys, and Peters [9]. They discovered that vitrified grinding wheels with a high modulus of elasticity wear by a mechanism of abrasive grain fracture rather than fracture of the glass bond bridges that hold the abrasive grains in place. As the modulus of elasticity increases the "self-sharpening effect" is lost because abrasive grains cannot be released from the bonding matrix. This leads to a condition where the temperature of the workpiece material begins to increase and is associated with phase transformations and thermal cracking of the surface layers that results in a reduction in fatigue strength.

In this case, the performance of the abrasive grinding wheel for a specific metal removal rate and workpiece material depends on the selection of the appropriate grade of abrasive grinding wheel that is a function of its modulus of elasticity and strength. In the present study, the elastic modulus, bending strength, and nature of fracture was found to be dependent on the vitrification behavior of the glass bonding system, the amount of bond, and the type of abrasive grain used in the vitrified grinding wheel. It was found that the wear of vitrified grinding wheels is highly dependent on the way the grinding wheel "vitrifies" during heat treatment.

2.3.2 Microstructure of Abrasive Grains

2.3.2.1 High Purity Aluminum Oxide

Examination of high purity aluminum oxide in a scanning electron microscope using an electron probe microanalyzer showed that 99.5 wt% of the grain was Al_2O_3 with the balance consisting of Na_2O and SiO_2 in equal proportions.

However, local Na_2O-enriched areas were observed within parts of the grain. Figure 2.17 shows the areas of Na_2O local enrichments within the grain as white

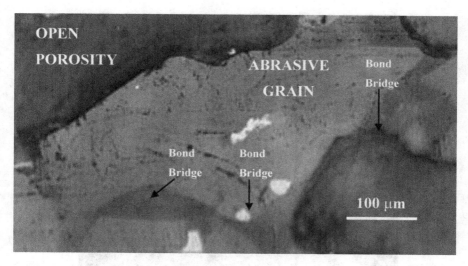

Fig. 2.17 High purity aluminum oxide grinding wheel showing enriched regions of Na_2O (denoted by *black arrows*) determined using an electron probe microanalyzer

reflections when viewed under an optical microscope. Under close examination, high purity aluminum oxide contains aluminum oxide, sodium aluminate, carnegie-ite, sodium monoaluminate, nepheline, and glass of variable composition. In heat-treated high purity abrasive grains, preferential etching at the surface of the grain appears to occur along crystallographically controlled directions (Fig. 2.18). This is assumed to be due to the dissolution of planar blocks of β-Al_2O_3 ($Na_2O \cdot 11Al_2O_3$) that is present in the α-Al_2O_3 host material. X-ray diffraction of high purity alumina established the existence of β-Al_2O_3 prior to the optical examination of the abrasive grains. Other impurities found include rarely seen calcium-rich platelets in the form of alite (Ca_3SiO_5), and an unnamed oxide, $NaCaAlO_3$, which is known to have several polymorphic forms.

2.3.2.2   Titanium-Doped Aluminum Oxide

The amount of TiO_2 in titanium-doped aluminum oxide was measured using an electron probe and was found to be in the range of 1–2 wt%. The amount of titania present is inconsistent with earlier work that had determined that the maximum solubility of TiO_2 in Al_2O_3 is less than 0.3 mol% at 1,300°C [79]. Although some of the excess can be accounted for in the formation of Ti_2O_3, it is possible that not all titania is in solid solution. This was confirmed by the occurrence of blade-like inclusions that is consistent with rutile (TiO_2) needle morphology. This would account for the variability in measured titania and its presence in amounts greater than its solubility in Al_2O_3. In heat-treated and titanium-doped aluminum oxide, calcium

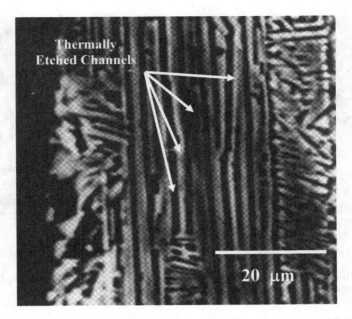

Fig. 2.18 High purity aluminum oxide grain showing thermally etched channels of β-Al_2O_3 layers that are present in the α-Al_2O_3 structure

hexaluminate, anorthite, and spinel are not affected by the heat treatment process. However, glass is devitrified forming anorthite spores. Titanium minerals are oxidized to higher oxides such as anatase and rutile. These changes are accompanied by large changes in volume that may affect the performance of any abrasive tool. As a precaution, Ti-doped aluminum oxide must be heated to 1,000°C before it can be used for making abrasive cutting tools.

2.3.2.3 Cubic Boron Nitride

Cubic boron nitride (CBN) abrasive grains are made by compacting grains of CBN in the presence of aluminum. Aluminum reacts with BN to form a mixture of AlN and AlB_2 that forms a stable and catalytically inactive binder. Interaction between aluminum and BN is intimate and can be observed directly using scanning and transmission electron microscopes. There is very little interaction between CBN grains. The edges of CBN grains not in contact with each other form rinds of AlN in thin, continuous lines with several nodules along its length.

The rind that encloses the exposed CBN grain is always orientated so that it has crystallographic directions parallel to particular directions in the CBN lattice. The selected area diffraction pattern shown in Fig. 2.19 shows a [110] CBN pattern with a rectangular AlN[11$\bar{2}$0] pattern superimposed. The AlN has grown with its basal

Fig. 2.19 (**a**) Two contacting CBN grains separated at intervals by an AlN rind which is parallel to the CBN [110] planes. The outer edges of the grains are in contact with a featureless AlB_2 layer. (**b**) Selected area diffraction pattern from part of the field of contact showing the relative orientation of phases present. The *smaller spots* are AlN, and the *larger spots* are CBN. The *arrow* indicates the single spot generated by AlB_2 phase

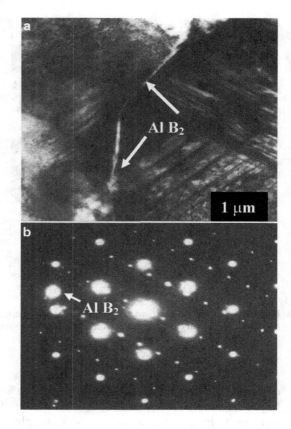

planes parallel to the CBN facet plane. This orientation with CBN (110)//AlN (0001) and cBN [110]// AlN [11$\bar{2}$0] is the most common orientation observed even when facet planes deviate away from being octahedral. At cube surfaces the orientation CBN (001)//AlN (0001) and CBN [110]// AlN[11$\bar{2}$0] occurs. While most of the AlN can be located at CBN grain surfaces, AlB_2 nucleates independently in liquid aluminum at the later stages of consolidation. A single crystal of AlB_2 produces the reflection to the left of the SAED pattern that produces a single bright spot [80].

2.3.3 Experimental Procedure

2.3.3.1 Measurement of Mechanical Properties

The experimental procedure involved making experimental samples of abrasive grain and glass bond as a vitrified product using high purity aluminum oxide, titanium-doped aluminum oxide, and CBN bonded with an alumino-borosilicate bond

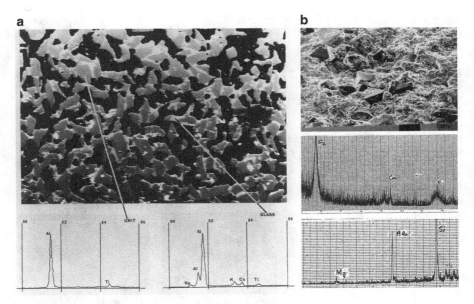

Fig. 2.20 Electron probe microanalysis of (**a**) titanium-doped aluminum oxide and vitrified glass bonding system, and (**b**) CBN and vitrified glass bonding system

containing 61.4 wt% SiO_2, 17 wt% Al_2O_3, 0.4 wt% Fe_2O_3, 3.2 wt% CaO, 0.1 wt% MgO, 2.7 wt% Na_2O, 3.1 wt% K_2O, and 10.1 wt% B_2O_3. Experimental samples were made by pressing abrasive grains and glass bond ingredients to a known density. The samples were moulded in the form of bars. The dimensions of the bars were 60 mm length, 12 mm height, and 12 mm depth. The samples were fired at the vitrification temperature (between 1,000 and 1,300°C) for 6 h in an electric furnace. The samples were prepared for four-point loading and for measuring their elastic modulus using the sonic method developed by [78]. A total of 20 experimental test samples were loaded in uniaxial tension. The Weibull modulus for the fractured samples was calculated to be 18.3 for aluminum oxide samples and 18.8 for CBN samples. A section of one of the bar samples was cut, mounted in resin, and polished to reveal the nature of bonding between glass and aluminum oxide. Figure 2.20 shows the section revealing abrasive grains bonded together by the vitrified glass bonding system. The black areas represent the pores between abrasive grains that are essential to provide free space for chips of metal and for coolant access. Figure 2.20 also shows the characteristic X-ray spectra for abrasive grains and glass bond. The abrasive grain spectrum shows aluminum and titanium (indicative of titanium-doped aluminum oxide), and the glass bond spectrum shows elements such as potassium, calcium, and sodium that are glass network-modifying elements, and aluminum and silicon that are network-forming elements. For the vitrified CBN grinding wheel, the bonding system contains magnesium, aluminum, silicon, calcium, and oxygen.

2.3.3.2   Manufacture of Grinding Wheels

Grinding wheel segments were made by pressing abrasive grains and glass bond ingredients to a known density. The samples were moulded in the form of segments to be attached to a pre-balanced grinding wheel body. The dimensions of the segments were 60 mm length, 15 mm height, and 20 mm depth. The samples were fired at the vitrification temperature (between 1,000 and 1,300°C) for 6 h in an electric furnace. Once fired, the segments were measured in terms of their hardness and grade and were bonded onto a steel backing using a high strength adhesive. The steel backings were then bolted onto a steel body containing the rest of the abrasive segments.

2.3.3.3   Measurement of Wear

The method of grinding wheel wear measurement adopted was the "razor blade" technique. The method involves grinding a workpiece that is less wide than the grinding wheel. A groove is worn into the wheel profile, which was measured with reference to the non-grinding portion of the grinding wheel using a razor blade. The grinding wheel was initially dressed using a single-point diamond and the wheel conditioned until steady-state grinding wheel wear was achieved. In order to achieve the conditions of bond fracture, the depth of cut for all experiments was set at 10 μm per pass with a table speed of 0.2 m/s and a grinding wheel speed of 60 m/s.

Immediately after the grinding experiments were performed, the razor blade was then lowered into the grinding position with the grinding wheel touching the blade. After grinding the blade, the wear of the grinding wheel was measured using a surface profilometer. The grinding ratio was calculated by measuring the volume of the grinding wheel removed and the volume of the workpiece removed.

2.3.4   Experimental Results

2.3.4.1   Mechanical Properties

The relationship between the elastic modulus and firing temperature as a function of abrasive grain type and bonding content is shown in Fig. 2.21 for both high purity and titanium-doped aluminum oxide structures. It is shown that the elastic modulus is developed as the vitrification temperature is increased and is highly dependent on the amount of bonding material that surrounds the abrasive grain. This is confirmed in Fig. 2.22, which shows the effect of the increase in bonding content on the elastic modulus at three different vitrification temperatures for high purity aluminum oxide structures. An interesting observation is that up to the softening point of the glass bond, high purity and titanium-doped aluminum oxide vitrified structures developed strength in the same way then declines for titanium-doped aluminum oxide structures depending on the amount of bonding material. The relationship is shown in Fig. 2.23. The same general trends are observed with vitrified CBN grinding wheels.

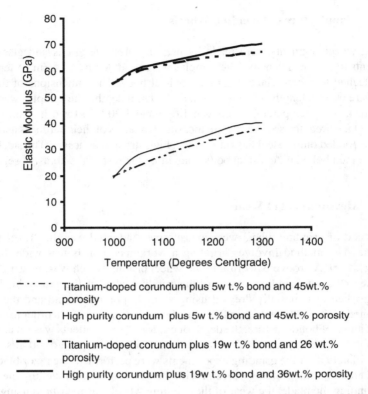

- - - - - Titanium-doped corundum plus 5w t.% bond and 45wt.% porosity

———— High purity corundum plus 5w t.% bond and 45wt.% porosity

- - - Titanium-doped corundum plus 19w t.% bond and 26 wt.% porosity

———— High purity corundum plus 19w t.% bond and 36wt.% porosity

Fig. 2.21 Elastic modulus as a function of firing temperature for a number of abrasive grain types and bond contents

Fig. 2.22 Effect of bond content and firing temperature on the elastic modulus of high purity aluminum oxide structures

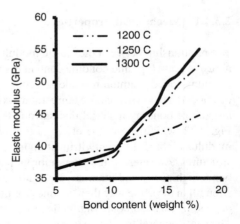

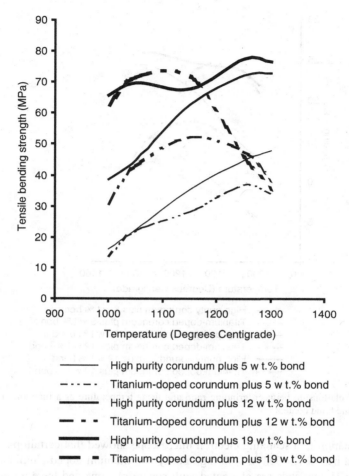

Fig. 2.23 Relationship between bending strength and firing temperature as a function of abrasive grain type and bond content

2.3.4.2 Wear of Grinding Wheels

The relationship between the wheel wear parameter, grinding ratio (G), and the firing temperature is shown in Fig. 2.24 for both high purity and titanium-doped aluminum oxide grinding wheel structures containing a different amount of vitrified bonding material. Again, the observation that up to the softening point of the glass bond, high purity and titanium-doped aluminum oxide structures develop wear resistance in the same way is noteworthy. Figure 2.24 shows that the grinding ratio is a function of vitrification temperature, but at a certain temperature, it is highly dependent on the type of abrasive grain used in the grinding wheel and the amount of bonding material used.

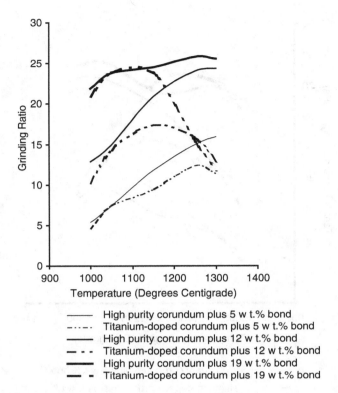

Fig. 2.24 Relationship between grinding ratio and firing temperature as a function of abrasive grain type and bond content

Examination in a scanning electron microscope showed that certain parts of the glass bond had devitrified in both high purity and titanium-doped aluminum oxide structures. The crystals are elongated with square sections and have a high Al_2O_3 content. An X-ray diffraction spectrum indicated that the phase is an aluminoborate solid solution. The best match was with $Al_{18}B_4O_{33}$. In addition to this phase, a second crystalline phase was observed in titanium-doped aluminum oxide structures. The phase consists of needles of rutile (TiO_2) orientated on the faces of titanium-doped aluminum oxide grains that penetrate into the glass bond. Figure 2.25a, b shows orientated rutile needle formation in the glass bond emanating from the aluminum oxide crystals. The structure in Fig. 2.25b was etched with a solution of 40% hydrofluoric acid in water. Figure 2.25c shows the growth of rutile needles from the interface between aluminum oxide and the glass bond using the electron backscatter mode. Figure 2.25d shows the devitrification of glass in the form of $Al_{18}B_4O_{33}$ crystals.

Fractured samples revealed a higher proportion of intergranular fracture than cut and polished samples. High purity aluminum oxide did not exhibit intergranular fracture at the interface between abrasive and bond but did exhibit the bond fracture mode. It appears that titania is an undesirable constituent in bonding systems that

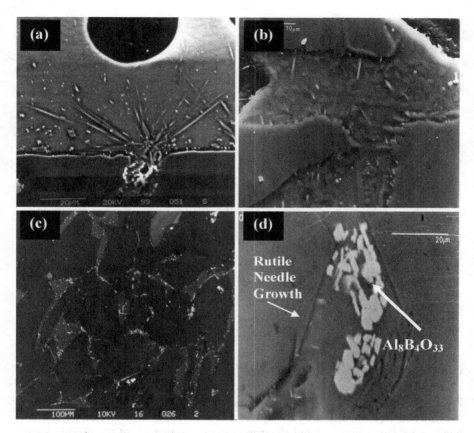

Fig. 2.25 (a) Titania (TiO_2), in the form of rutile needles, on the surface of the vitrified glass bond; (b) vitrified glass bond etched with 40% HF in water to show rutile formation within the glass bonding system; (c) electron backscattered image showing needle growth into the glass bond from the abrasive; (d) devitrified glass bond containing crystals of $Al_{18}B_4O_{33}$ bounded by two abrasive grains

tends to promote interfacial fracture at the abrasive grain-bond bridge interface. Even if its presence does not cause a reduction in cohesive strength, one method of reducing its effect is for it to form a titanate compound that does not reduce interfacial strength. Examination of fractured high purity aluminum oxide samples revealed preferential etching of the abrasive grain by the glass bond. This is assumed to be dissolution of blocks of β-aluminum oxide ($Na_2O \cdot 11Al_2O_3$) present in α-aluminum oxide (pure aluminum oxide). The relationship between the wheel wear parameter, grinding ratio, and the firing temperature for vitrified CBN grinding wheel structures containing different amounts of bonding content is shown in Fig. 2.26. An interesting observation one can observe is that the retention of the abrasive grains in the vitrified bonding matrix can be improved by increasing the sintering temperature. In order to investigate the mechanism of CBN retention, samples of the post-fired abrasive structures were polished and etched. Figure 2.27 shows the

Fig. 2.26 Relationship between grinding ratio and firing temperature as a function of bond content for vitrified CBN grinding wheel structures

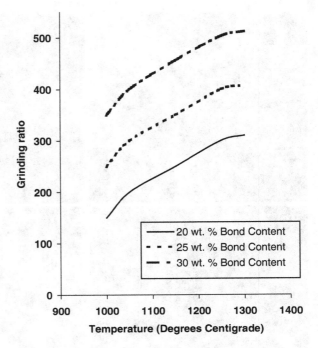

unpolished fracture surfaces of the vitrified CBN grinding wheels. A magnified image of the interface between abrasive grain and bonding bridge is shown in Fig. 2.27b. Interfacial cohesion appears to be quite apparent in this image. Figure 2.28 shows a polished and etched fracture surface in the vicinity of the abrasive grain and bonding bridge. The associated electron probe microanalysis of the image clearly shows a concentration of oxygen at the interface between CBN and glass bonding bridge. The concentration of oxygen appears to be associated with boron and the formation of a boron-containing oxygen layer that separates the alumino-borosilicate bonding system and the CBN abrasive grain. This is thought to be a relatively thin layer of B_2O_3 (boric oxide). As the sintering temperature is increased, the thickness of this layer is also increased with a subsequent loss of boron from the abrasive grain. Figure 2.29 illustrates the relationship between the interfacial layer thickness and sintering temperature. As the temperature is increased further, the width of the interfacial layer tends to stabilize and reaches an equilibrium thickness.

2.3.5 Discussion of Interfacial Compounds on Grinding Wheel Wear

The existence of β-aluminum oxide was established by X-ray methods. When the bond content is low in samples made with high purity aluminum oxide, failure

Fig. 2.27 (a) Vitrified CBN grinding wheel structure, (b) interface between CBN abrasive grain and vitrified bonding

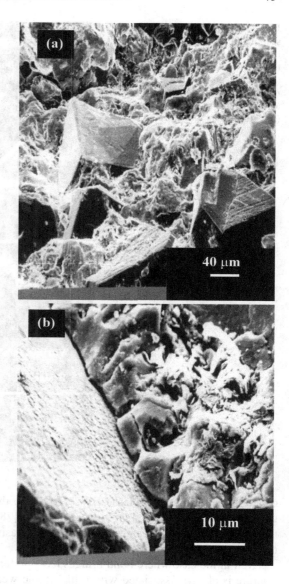

occurs by fracture of bonding bridges. At higher bond contents the mode of failure is one of abrasive grain fracture. Fracture at the abrasive grain-bond bridge interface was not observed. This is because the β-aluminum oxide phase is etched away preferentially due to the dissolution of Na_2O into the glass bond that locally increases the fluidity of the bond. This allows the bond to penetrate the surface of the abrasive grain and provides it with enhanced shear resistance.

This effect does not happen with titanium-doped aluminum oxide, in fact, the strength decreases at the softening point of the glass because of enhanced dissolution

Fig. 2.28 (a) Polished cross section of CBN abrasive and bond bridge clearly showing the interface layer, (b) electron probe microanalysis of oxygen across the line scan shown in (a), left-to-right

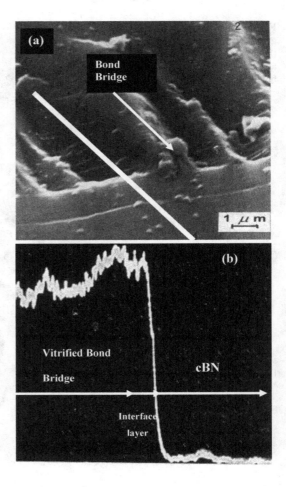

of aluminum oxide that releases more TiO_2 into the glass bond for rutile needle growth. Therefore, in contrast to Decneut et al. [78], the mode of fracture in titanium-doped structures is interfacial between abrasive grain and glass bond and is not completely dependent on bond content.

Even in the case where bond bridges have preferentially fractured, the mode of fracture is always associated with rutile needle weakening. The vitrification temperature and glass bond content has a significant effect on the elastic modulus of high purity and titanium-doped aluminum oxide structures. The differences in strength between these structures when fired at temperatures above the softening point of the glass bond are due to differences in the crystal structures of the two types of abrasive grain. The presence of β-aluminum oxide in high purity aluminum oxide allows selective dissolution of aluminum oxide to occur that enables stronger bonding to take place between aluminum oxide and glass. This effect does not happen with titanium-doped aluminum oxide where dissolution allows the precipitation

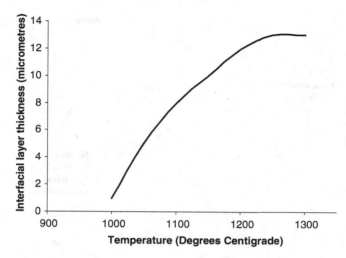

Fig. 2.29 Interfacial layer thickness between CBN and vitrified bonding bridge as a function of sintering temperature

of TiO_2 into the glass bond in the form of rutile needles that reduces the cohesive strength between aluminum oxide and glass.

The existence of an interfacial layer between CBN and glass was thought to be that of the formation of boric oxide (B_2O_3). As sintering continued, the layer became thicker and tended to strengthen the interfacial layer. This is assumed to be the reason why the grinding ratio of the abrasive tool increased as a function of sintering temperature. It was also noted that the size of the CBN grains decreased as sintering temperature increased until an equilibrium interfacial layer thickness was reached. It would also be right to assume that at this point that diffusion of oxygen into the CBN abrasive grain ceases to occur. The fracture surface of the vitrified CBN structure shows that fracture is associated with fracture within the bonding bridge rather than fracture at the CBN–bond bridge interface. This tends to imply that the interfacial bonding layer is stronger than bonding bridge.

2.4   Case Study II: Dissolution of Quartz and Its Effect on Grinding Wheel Wear

When considering individual bond constituents, mineral fluxes and ground glass frits have little direct effect on the ability to manufacture grinding wheels. However, most clay minerals develop some plasticity in the presence of water, which improves the ability to mould the mixture so that the wheel, in its green state, can be mechanically handled [80, 81]. Clays and clay-based fluxes contain an amount of free quartz that has a detrimental effect on the development of strength during vitrification heat treatment. Clays are used to provide vitrified grinding wheels with green strength

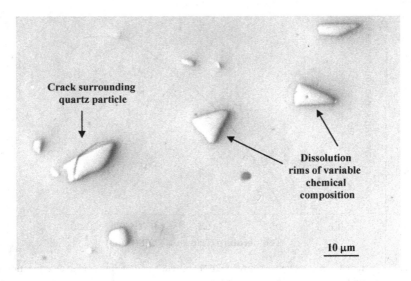

Fig. 2.30 A collection of quartz particles in a vitrified bonding system. The quartz particle on the left has a circumferential crack extending into the dissolution rim and abrasive grain

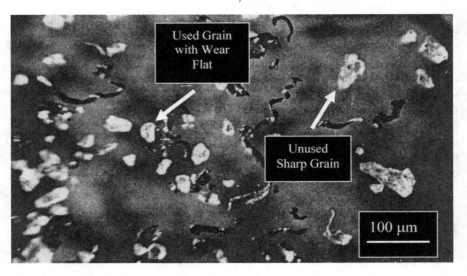

Fig. 2.31 Grinding swarf and a collection of used and unused abrasive cutting grains

during the heat treatment process. However, when the glass material solidifies around the particles of clay and quartz, the displacive transformation of quartz during the cooling stage of vitrification leads to the formation of cracks in the glass around the quartz particle (Fig. 2.30). The strength of the bonding bridge is reduced and leads to the early release of the abrasive particle during the cutting process (Fig. 2.31).

The basic wear mechanisms that affect vitrified grinding wheels are concerned with grain fracture during metal cutting, fracture of bond bridges, mechanical fracture of abrasive grains due to spalling, and fracture at the interface between abrasive grain and bond bridge [82–96]. Failure in vitrified silicon carbide grinding wheels is more probable due to the lack of a well-developed bonding layer between abrasive grain and the glass bond bridge, which is typically only a few micrometers. The lack of adequate bonding is due to the use of a high clay content bonding system with very little opportunity for a glass to form at the interface. High glass content bonding systems tend to aggressively decompose the surface of silicon carbide abrasive grains. In vitrified aluminum oxide grinding wheels, high glass content bonding systems are used extensively and lead to bonding layers in excess of 100 μm in thickness.

In addition to the formation of very thin bonding layers in vitrified silicon carbide grinding wheels, the use of high clay content bonding systems means that there is an increase in the amount of quartz contained in the bond bridges between the abrasive grains. Although the likelihood of decomposition of silicon carbide surfaces is reduced, the probability of bond bridge failure is increased due to the increased quartz content. Therefore, the dissolution of quartz in these bonds is highly desired in order to compensate for a much thinner interfacial bonding layer. Difficulties encountered when developing a dissolution model arise from the fact that the phase boundary between quartz particle and molten glass moves during the diffusion process. The problem of a fixed boundary can be solved without difficulty although this is not equivalent to the conditions associated with a moving boundary between quartz particle and a highly viscous glass melt. The development of dissolution models is required to determine the magnitude of quartz remaining in the bonding system after a period of heat treatment. The models are then compared with experimentally determined quartz content of the bonding systems using X-ray diffraction techniques. Subsequently, dissolution models are used to specify the appropriate heat treatment schedule for a particular bonding system that is used in grinding wheels that grind automotive camshafts and crankshafts depending on the material removal rate and the nature of the material to be ground. The use of X-ray techniques is also applied to measuring phase transformations in grinding wheels that have been subjected to laser irradiation. When using a laser beam to dress the wheel, it is possible to form localized texture in the abrasive grains that allow the grinding wheel to remove material in the superfinishing regime. For the first time, it is reported that grinding wheels are able to provide roughing, finishing, and superfinishing operations in one grinding stroke.

2.4.1   Dissolution Models for Vitrified Grinding Wheel Bonds

When densification occurs in a vitrified grinding wheel after the peak soaking temperature has been reached, the cooling rate is reduced to prevent thermal stress cracking in the bonding layers between abrasive grains. Cooling rates are reduced when crystalline inversions occur that involve volume changes. The inversion range for quartz and cristobalite is 550–580°C and 200–300°C, respectively. Since the formation of cristobalite is rare in vitrified bonding systems used for grinding wheels, the rapid displacive transformation of quartz tends to promote the formation of cracks in bonding bridges.

When quartz-containing bonds begin to cool from the soaking, or vitrification, temperature it is thought that the liquid phase relieves stresses resulting from the thermal expansion mismatch between itself and the phases, β-quartz, β-cristobalite, and mullite, to at least 800°C. At 800°C, stresses will develop in quartz particles and in the matrix that causes cracking to occur around quartz particles. The shrinkage behavior of quartz and the glass phase has been described by Storch et al. [97]. Between the temperature range, 573 and 800°C, the glass phase shrinks more than the quartz phase that causes tangential tensile stresses to form cracks in the matrix. At 573°C, β-quartz transforms to α-quartz that causes residual stresses around quartz particles to produce circumferential cracking around those same quartz particles (Fig. 2.30). Some of these cracks have been seen to propagate into the glass phase [98]. Similar observations occur in the cristobalite phase. Spontaneous cracking of quartz has been found to occur over a temperature range that depends on the size of the quartz particles [99]. Particles larger than 600 μm diameter cracked spontaneously at 640°C, whereas smaller particles of less than 40 μm diameter cracked at 573°C. This observation agrees with temperature-dependent cracking reported by Kirchhoff et al. [100]. To maintain the integrity of the bond bridges containing coarse quartz particles, the grinding wheel must remain at the vitrification temperature until the quartz particles have dissolved.

The dissolution model assumes that at a constant absolute temperature, T, a particle of quartz melts in the surrounding viscous glass melt and that the rate of change of the volume of quartz present in the melt at a particular instant in time is proportional to the residual volume of quartz. The above assumption is based on the fact that alkali ions diffuse from the viscous glass melt to the boundary of the quartz particle thus producing a dissolution rim around each quartz particle. Diffusion rims around quartz particles are shown in Fig. 2.30.

A high reaction rate will initially occur which continuously decreases as the quartz particle is converted to a viscous melt. Previous models have provided an insight into how various factors contribute to the dissolution of quartz in vitreous bodies. However, Jackson and Mills [101] derived a mathematical relationship that accounts for the change in density when β-quartz transforms to α-quartz on cooling from the vitrification temperature, thus,

$$m_{T,t} = M\gamma \exp\left(-At^{1/2} \exp\left[\frac{-B}{T}\right]\right), \qquad (2.2)$$

where, $m_{T,t}$ is the residual mass fraction of quartz at a constant time and temperature couple, M is the original mass fraction of quartz prior to heat treatment, γ is the ratio of densities of β-quartz and α-quartz, A and B are constants, t is time, and T is

Table 2.3 Chemical analyses of raw materials

Oxide (wt%)	China clay	Ball clay	Potash feldspar	Quartz
Al_2O_3	37	31	18.01	0.65
SiO_2	48	52	66.6	98.4
K_2O	1.65	1.8	11.01	0.35
Na_2O	0.1	0.2	3.2	0.04
CaO	0.07	0.2	0.09	0.00
MgO	0.03	0.3	0.09	0.00
TiO_2	0.02	0.9	0.00	0.07
Fe_2O_3	0.68	1.1	0.11	0.03
Loss on ignition	12.5	16.5	0.89	0.20

Table 2.4 Mineralogical analyses of raw materials

Compound (wt%)	China clay	Ball clay	Potash feldspar	Quartz
Quartz	4.05	12.77	4.93	98.40
Orthoclase	0.00	15.23	64.96	0.00
Kaolinite	79.70	62.71	2.17	0.00
Mica	13.94	0.00	0.00	0.00
Soda feldspar	0.8	1.69	27.07	0.00
Miscellaneous oxides/losses	1.51	7.60	0.87	1.60

absolute temperature. The model was compared with experimental data determined using the powder X-ray diffraction method. The experimental work was divided into two parts. The first part concentrates on comparing the dissolution model with X-ray diffraction data using "sintering" bond compositions that are used in vitrified silicon carbide grinding wheels, whilst the second part focuses on comparing the model with "fusible" bond compositions that are used in high-performance vitrified aluminum oxide grinding wheels.

2.4.2   Experimental Procedures

2.4.2.1   Raw Materials and Preparation

The raw materials used in the experimental study (case study 2) were Hymod Prima ball clay, standard porcelain China clay, potash feldspar, and synthetic quartz (supplied as silica flour). The chemical analysis of the raw materials is shown in Table 2.3. Rational analysis of the raw materials was performed to reveal the mineralogical composition of the raw materials. The rational analysis appears in Table 2.4. The bond mixture described is one typically used in vitrified silicon carbide grinding wheels where the erosion of the abrasive grain is reduced by using high clay content bonding systems. This bonding system is used where silicon carbide is predominantly used in grinding cast iron camshafts and crankshafts.

Fusible bonding systems using a mixture of ball clay and potassium-rich feldspar were made to test the model developed by Jackson and Mills [101]. The ball clay used contained 12.77 wt% quartz, and the feldspar contained 4.93 wt% quartz. The bonding system was composed of 66 wt% ball clay and 34% feldspar. The initial quartz content, M, of the bond mixture was 10.1 wt%. The bond mixture described is one typically used in high-performance vitrified aluminum oxide grinding wheels and is used when grinding steel camshafts and crankshafts.

The raw materials were mixed in a mortar, pressed in a mould, and fired at various temperatures. A heating rate of 3°C/min was employed until the vitrification temperature was reached. The typical soaking temperature was varied between 1,200 and 1,400°C for "sintering" bond compositions, and 950 and 1,050°C for "fusible" bond compositions in order to simulate industrial firing conditions. The samples were cooled at a rate of 2°C/min to avoid thermal stress fracture in the bonding bridges between abrasive grains. The fired samples were crushed to form a fine powder in preparation for X-ray diffraction.

2.4.2.2   X-ray Diffraction of Vitrified Bonding Systems

The dissolution model was compared with experimental data using the X-ray powder diffraction method. X-ray diffraction of the raw materials was performed on a Phillips 1710 X-ray generator with a 40 kV tube voltage and a 30 mA current. Monochromatic Cu kα radiation, $\lambda = 0.154060$ nm, was employed. A scanning speed of 2° per minute for diffraction angles of 2θ was used between 2θ angles of 15° and 60°, and the X-ray intensity was recorded using a computer. The spectrum was then analyzed and compared with known spectra. Powder specimens were prepared by crushing in a mortar and pestle in preparation for quantitative X-ray diffraction. To eliminate the requirement of knowing mass absorption coefficients of ceramic samples for quantitative X-ray diffraction, Alexander and Klug [102] introduced the use of an internal standard. Firstly, the ceramic sample is crushed to form a powder—the sizes of particles should be small enough to make extinction and absorption effects negligible. Secondly, the internal standard to be added should have a mass absorption coefficient at a radiation wavelength such that intensity peaks from the phase(s) being measured are not diminished or amplified. It should be noted that the powder diffraction mixture should be homogeneous on a scale of size smaller than the amount of material exposed to the X-ray beam and was free from preferred orientation. The powder bed that is subjected to "X-rays" should be deep enough to give the maximum diffracted intensity. The expected equilibrium phases from the fired mixtures are quartz (unreacted and partially dissolved), mullite, cristobalite, and glass. However, from the samples tested, the compounds quartz, mullite, and glass were successfully detected. A calibration curve was constructed using a suitable internal standard (calcium fluoride), a diluent, and a synthetic form of the phase(s) to be measured. Synthetic mullite had a purity greater than 99.8%, whilst powdered quartz had a purity greater than 99.84% SiO_2. The method used for quantitative analysis of ceramic powders was developed by Khandelwal and Cook [103].

Fig. 2.32 Calibration curve for quantitative analysis of X-ray determined quartz and mullite using the CaF$_2$ (111) plane generated by the internal standard

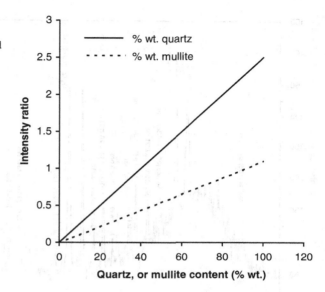

The internal standard provides an intense (111) reflection (d=0.1354 nm) lying between the (100) reflection for quartz (d=0.4257 nm) and the (200) reflection for mullite (d=0.3773 nm). Using Cu kα radiation (λ=0.15405 nm), the corresponding values of diffraction angle 2θ are: (100) quartz=20.82°; (111) calcium fluoride=28.3°; and (200) mullite=32.26°. Figure 2.32 shows the calibration curve generated by varying proportions of calcium fluoride, synthetic quartz, and mullite. Mass fractions of the crystalline phases in the mixture can be interpreted from the calibration lines by measuring the intensity ratio of the phase(s) to the internal standard. Figure 2.33 shows the diffraction peaks of interest for quantitative analysis lying between 15° and 40° of the diffraction angle 2θ. The figure shows the reflections of the (111) plane of calcium fluoride, (200) plane of mullite, and the (100) plane of quartz. In order to calculate the mass fractions of quartz and mullite in the mixture, the height of the chosen diffraction peak and its width at half-height were measured from the diffraction spectrum. The product of these two measures were then compared with that of the internal standard, and the resultant intensity ratio was used to find the exact mass fraction of the phase(s) measured in the glass that was subjected to X-ray diffraction.

2.4.2.3   Grinding Wheel Performance

A series of grinding wheel experiments were conducted in order to show the difference between bonding systems with different levels of quartz content contained in their bonding bridges. The experiments were conducted using high-speed steels and a high chromium content hypereutectoid steel (AISI 52100) in order to compare with field trials conducted using commercially available vitrified aluminum oxide

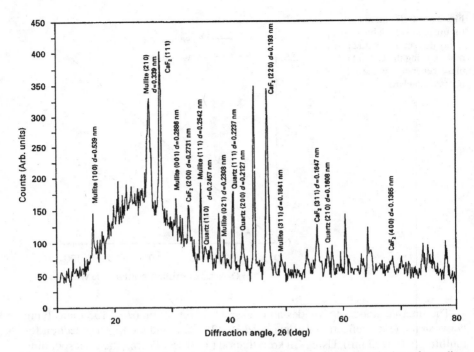

Fig. 2.33 X-ray diffraction spectrum of a vitrified bonding system showing the interplanar distances of crystallographic planes of mullite, quartz, and calcium fluoride. Scan rate was 2° per minute

grinding wheels. A series of controlled experiments were designed to compare grinding wheels under increasing rates of metal removal. The experiments were terminated when a condition of severe burn, chatter, or wheel breakdown was observed. The initial experimental wheels used were: angular white alumina with a low temperature bonding system (wheel specification A); a sol-gel alumina abrasive wheel with angular white abrasive mixed in a one-to-one proportion bonded with a low temperature fired bonding system (wheel specification B); and a monocrystalline alumina wheel with a low temperature fired bonding system (wheel specification C). All wheels were manufactured with a vitreous bond, 60 mesh size abrasive grain (approximately 220 μm in diameter) material, J-hardness grade, and a fairly open structure. Experiments were performed on a Jones and Shipman series ten cylindrical grinding machine using a 450 mm diameter grinding wheel rotating at 33 m/s surface speed. The wheel was dressed using a diamond blade tool using a depth of cut of 30 μm at a feed rate of 0.15 mm/rev, and a final dressing depth of cut of 15 μm prior to grinding workpieces. The amount of material removed was 250 μm per grinding stroke. The coolant flow pressure was 0.5 bar at a flow rate of 15 l/min using a 2% concentrated solution of oil in water.

2.4.3   Experimental Results

2.4.3.1   Silicon Carbide Bonding Systems—Verification and Comparison of Dissolution Models for Quartz

In addition to comparing the experimental results to the dissolution model, results published in the literature were also used to test the accuracy of the model. The composition of the experimental mixtures was matched to those specified by Lundin [81]. Lundin's experimental mixtures were composed of 25 wt% quartz (13.2 μm particle size), 50 wt% clay (kaolin), and 25 wt% flux (potassium feldspar −25 μm average particle size).

The constants A and B for the sintering bonding system were calculated as

$$A = 5.62 \times 10^8, \tag{2.3}$$

$$B = 33374. \tag{2.4}$$

From which the experimental activation energy, Q, is 132.65 kcal/mol. The residual quartz content for the sintering bonding system is given as follows:

$$m_{T,t} = 26.25 \cdot \exp\left[-5.62 \times 10^8 \cdot t^{1/2} \cdot e^{-33,374/T}\right]. \tag{2.5}$$

The data comparing Lundin's experimental results, the author's experimental results, and the dissolution model is shown in Table 2.5. When the data is plotted as the logarithm of $(-\ln[m/M]/t^{1/2})$ versus the reciprocal of absolute temperature, $1/T$, then all data fits a straight-line relationship. The gradient was calculated to be 33,374, the constant B, using two data points. Lundin's experimental gradient gave a value of 32,962 using the least squares method, and 34,000 for the present work. The corresponding activation energies for both systems are 131 kcal/mol for Lundin's work [81] and 135 kcal/mol for the present work, respectively. Figures 2.34 and 2.35 show the effects of time on residual quartz content at different temperatures according to (2.5) together with comparative experimental data.

A comparison was made with dissolution models published in the literature. One of the earliest models was derived by Jander [104]. The equation can be expressed as

$$\left(1 - \sqrt[3]{1-Z}\right)^2 = \left\{\frac{C_1 \cdot D}{r^2}\right\} \cdot t, \tag{2.6}$$

where Z is the volume of quartz that has been dissolved, r is the original particle radius, and D is the diffusion coefficient for the diffusing species. This equation can be transformed into mass fractions using Archimedes' law, thus,

$$\left(1 - \sqrt[3]{\frac{m}{M}}\right)^2 = C_2 \cdot t, \tag{2.7}$$

Table 2.5 Residual quartz content of a sintering bonding system at various vitrification temperatures

Temperature (°C)	Time (h)	Lundin's experimental result (wt%)	Experimental result (wt%)	Jackson and Mills' [101] result (wt%)
1,200 (1,473 K)	1	24.1	24.2	24.2
1,200	1	24.7	24.3	24.2
1,200	1	26.1	24.8	24.2
1,200	2	23.7	23.8	23.4
1,200	2	23.6	23.9	23.4
1,200*	2	23.4	23.4	23.4
1,200	4	21.3	22.2	22.3
1,200	8	20.3	20.9	20.8
1,200	18	19.0	18.5	18.6
1,200	18	18.9	18.6	18.6
1,200	48	15.2	15.1	14.9
1,250 (1,523 K)	1	22.7	22	22.1
1,250*	2	20.6	20.6	20.6
1,250	4	18	18.5	18.6
1,250	8	15.5	16	16.2
1,250	18	12.6	12.5	12.6
1,250	48	8.3	7.8	8.0
1,300 (1,573 K)	0.5	22.6	20.4	20.6
1,300	0.5	21	20.9	20.6
1,300	1	20	18.3	18.6
1,300	2	16.1	15.9	16.2
1,300	4	13.4	12.8	13.2
1,300	8	10	9.7	9.9
1,300	18	5.9	5.8	6.1
1,300	50	1.6	1.8	2.3
1,300	120	0.3	0.2	0.6

Lundin's [81] experimental data is compared with the author's experimental data and the model [101]. The *asterisk* indicates that the values used for deriving the constants are used in the theoretical model

where, C is a constant dependent on soaking temperature and initial particle size of quartz. Krause and Keetman [105] expressed the dissolution of quartz as a function of isothermal firing time, viz,

$$M - m = C_3 \cdot \ln t,\qquad(2.8)$$

where M is the initial quartz content, m is the residual quartz content after time, t. The unit of time here is seconds such that after 1 s of firing the residual quartz content is equal to the initial quartz content. Monshi's dissolution model [106] can be transformed into the following equation assuming isothermal firing conditions

$$\ln\left\{\frac{m}{M}\right\} = -C_6\sqrt{t}.\qquad(2.9)$$

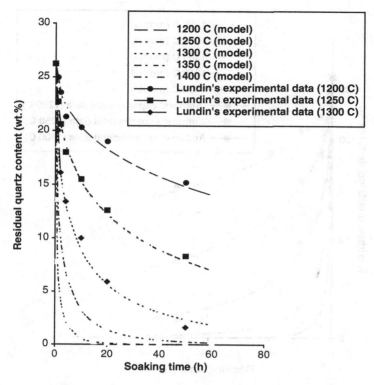

Fig. 2.34 Effect of time on residual quartz content of a sintering bonding system according to Jackson and Mills model [101] and compared with Lundin's experimental data [81]

Jackson and Mills' model [108] for isothermal firing conditions is transformed into

$$\ln\left\{\frac{m}{\gamma . M}\right\} = -C_7\sqrt{t}, \qquad (2.10)$$

where γ is the ratio of densities of β- and α-quartz. Constants for all the equations presented here are calculated using quartz mass fraction data after 18 h firing. The constants are dimensioned in seconds. The equations shown were compared with experimental data generated by Lundin [81] for a clay-based material containing 40 wt% kaolin, 40 wt% quartz, and 20 wt% feldspar. According to the transformed equations, the mass fraction of quartz can be calculated as follows:

Jander's model [104]

$$m = 41.9 \cdot \left(1 - \left\{1.55 \times 10^{-6} \cdot t\right\}\right)^{3/2}. \qquad (2.11)$$

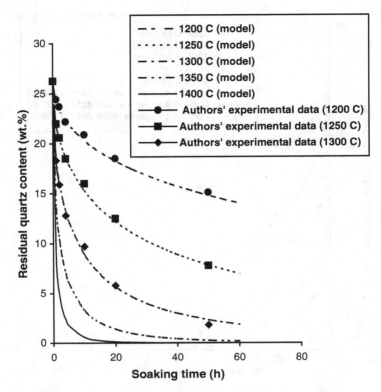

Fig. 2.35 Effect of time on residual quartz content of a sintering bonding system according to Jackson and Mills' model [101] and compared with the authors' experimental data

Krause and Keetman's model [105]

$$m = 41.9 - (2.58 \cdot \ln t) \tag{2.12}$$

Monshi's model [106]

$$m = 41.9 \cdot e^{-4.5 \times 10^{-3} \sqrt{t}}. \tag{2.13}$$

Jackson and Mills' model [101]

$$m = 41.73 \cdot e^{-4.5 \times 10^{-3} \sqrt{t}}. \tag{2.14}$$

The transformed equations are then tested using data provided by Lundin [81]. Referring to Table 2.6, it can be shown that the mass fraction of quartz obtained using the equations derived by Jander [104] and Krause and Keetman [105] did not agree with Lundin's experimental results [81]. The results obtained using Monshi's model [106] are in much better agreement compared to Lundin's data.

However, the results obtained using Jackson and Mills' model [101] are more accurate at predicting the mass fraction of quartz remaining owing to the differences

Table 2.6 Residual quartz content for different soaking times at 1,300°C for a sintering bonding system composed of 40 wt% kaolin, 40 wt% quartz, and 20 wt% feldspar (Lundin's mixture number M21 [81]) compared with other dissolution models

Time (h)	Lundin's experimental data [81]	Jander [104]	Krause and Keetman [105]	Monshi [106]	Jackson and Mills [101]
0	41.9	41.9	0.00	41.9	41.9
0.5	35.9	41.72	22.55	34.61	34.76
1	32.8	41.54	20.76	31.97	32.12
2	29.2	41.19	18.97	28.58	28.72
4	23.2	40.49	17.18	24.39	24.51
8	19.5	39.11	15.39	19.49	19.59
18	13.3	35.72	13.30	13.30	13.36
24	10.7	33.74	12.56	11.13	11.19
48	6.9	26.18	10.77	6.43	6.51
120	3.6	7.85	8.96	2.17	2.17
190	2.7	0.00	7.22	1.00	1.01
258	2.0	0.00	6.43	0.54	0.55

in the density of quartz. After long periods of heat treatment, the model predicts lower magnitudes of mass fractions of quartz when compared to Lundin's experimental results [81].

2.4.3.2   Aluminum Oxide Bonding Systems—Verification and Comparison of Dissolution Models for Quartz

The constants, A and B, for the fusible bonding system were determined using time and temperature couples at 2 and 10 h and were calculated to be, -5.2×10^8 and $-33,205$, respectively. The dissolution equation then becomes

$$m_{T,t} = 10.06 \exp\left[-5.2 \times 10^8 t^{1/2} \cdot e^{-33,205/T}\right]. \tag{2.15}$$

Equation (2.15) is used to compare the experimentally determined mass fraction of quartz remaining after heat treatment with the predicted values. The calculated mass fraction of quartz remaining after a period of heat treatment is calculated using the equation derived by Jackson and Mills [101]. The results of the dissolution model compare well with the experimental data over short periods of time. However, over longer periods of heat treatment the model tends to become less accurate (Fig. 2.36). A comparison was made with published dissolution models. The equations shown were compared with experimental data at 1,050°C. According to the transformed equations, the mass fraction of quartz can be calculated as follows:

Jander's model [104]

$$m = 10.1 \cdot \left(1 - \left\{3.44 \times 10^{-6} \cdot t\right\}\right)^{3/2}. \tag{2.16}$$

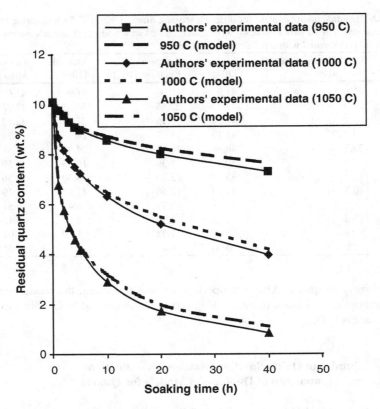

Fig. 2.36 Effect of time on residual quartz content of a fusible bonding system according to Jackson and Mills' model [81] and compared with the author's experimental data

Krause and Keetman's model [105]

$$m = 10.1 - (0.59 \cdot \ln t). \tag{2.17}$$

Monshi's model [106]

$$m = 10.1 \cdot e^{-6.4 \times 10^{-3} \sqrt{t}}. \tag{2.18}$$

Jackson and Mills' model [81]

$$m = 10.06 \cdot e^{-6.37 \times 10^{-3} \sqrt{t}}. \tag{2.19}$$

Table 2.7 Residual quartz content for different soaking times at 1,050°C for a fusible bonding system compared with other dissolution models

Time (h)	Experimental data	Jander model [104]	Krause and Keetman model [105]	Monshi's model [106]	Jackson and Mills' model [81]
0	10.1	10.1	0	10.1	10.1
1	6.84	9.91	5.23	6.88	6.86
2	5.79	9.72	4.82	5.87	5.86
3	5.13	9.54	4.58	5.21	5.19
4	4.7	9.36	4.41	4.7	4.68
5	4.28	9.18	4.28	4.28	4.28
10	3.2	8.28	3.87	2.99	3
20	2	6.6	3.46	1.81	1.82
40	1.1	3.62	3.04	0.89	0.89

With reference to Table 2.7, it can be shown that the mass fraction of quartz obtained using the equations derived by Jander [104] and Krause and Keetman [105] did not agree with the experimental results at 1,050°C. The results obtained using Monshi's model [106] are in much better agreement compared to the experimental data. However, the results obtained from Jackson and Mills' model are more accurate at predicting the mass fraction of quartz remaining owing to the differences in the density of quartz. After long periods of heat treatment, the model predicts slightly lower magnitudes of mass fractions of quartz when compared to the experimental results. The use of X-rays to predict the level of quartz in vitrified bonding systems can be used to specifically design grinding wheels for specific grinding processes where the quartz content in the bonding system will reduce the economic impact of using vitrified alumina grinding wheels.

2.4.3.3 Grinding Wheel Experiments

The dissolution of quartz during heat treatment has a significant effect on the wear of vitrified grinding wheels. Figure 2.37 shows the effect of using a high- and a low-quartz content bonding system on the wear of vitrified aluminum oxide grinding wheels grinding a large number of tool steel materials [107]. The classification of tool steels is in the form of an abrasive hardness number, which is a weighted average of the number of carbides contained within the tool material. As shown in Fig. 2.37, the grinding ratio, or G-ratio, is a measure of the efficiency of the grinding wheel. It is the quotient of the volume of workpiece material removed and the volume of the wheel material removed. The figure demonstrates the effectiveness of reducing the quartz content of the bonding system. X-ray diffraction techniques have been used to characterize the bonding system and is an effective method in the selection of raw materials used for high efficiency grinding wheels.

Further experimental results using hypereutectoid steels were compared with field experiments using camshaft and crankshaft grinding operations as the basis for comparison. Forty workpiece samples were ground and the results of the initial

Fig. 2.37 Effect of the abrasive number on the grinding ratio for a high-quartz content and a low-quartz content bonding system grinding tool steel materials in the cylindrical surface grinding mode. Tool steel material is marked on the trend lines and is a function of the carbide content in their microstructure expressed as an abrasive number

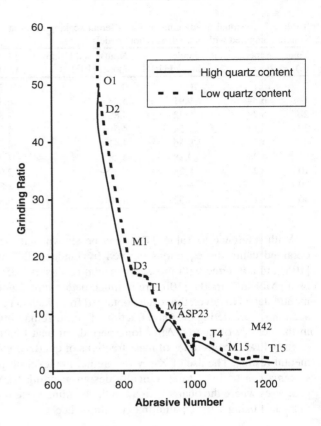

experiments are shown in Figs. 2.38, 2.39, and 2.40. None of the wheels used produced chatter vibrations or burned the workpieces, which was represented by a change in the level of grinding power. The grinding wheels gave a similar performance level, surface roughness, and grinding ratio. At low grinding rates, it was possible to differentiate between the grinding performance of each wheel.

The sol-gel wheel gave the best performance in terms of surface roughness and the highest grinding ratio whilst the angular white alumina wheel provided the worst results that resulted in rapid wheel wear. The use of monocrystalline abrasives at moderate metal removal rates ground approximately 30 workpiece materials before it started to breakdown. From these initial results, it can be shown that sol-gel abrasive wheels operate well at high power levels and can grind more efficiently at higher metal removal rates compared to angular white and monocrystalline abrasive grinding wheels. At low metal removal rates, there appears to be no significant difference between the grinding wheels used in the experiments. These experiments also show that bonding systems that are rich or depleted in quartz particles have a significant effect on the breakdown of the grinding wheel during grinding experiments—a significant factor in the subsequent change in grinding ratio.

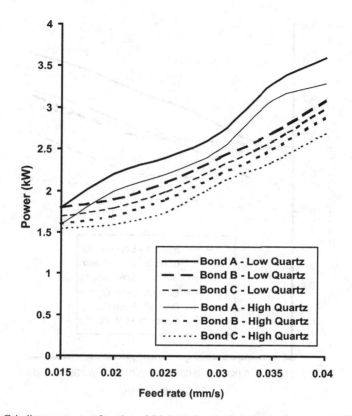

Fig. 2.38 Grinding power as a function of feed rate for vitrified aluminum oxide grinding wheels with different types of abrasive and bond. Grinding conditions: wheel surface speed = 33 m/s; depth of cut = 0.25 mm; dress conditions = diamond blade traversed at 0.15 mm/rev at a depth of cut of 0.03 mm

2.5 Discussion

A model describing the dissolution of quartz has been used to predict the mass fraction of quartz remaining after a period of heat treatment in vitrified grinding wheel bonding systems. The model has been compared with experimentally determined mass fractions of quartz in an industrial vitrified bonding system. However, the assumptions made when formulating the model may invalidate its wider application. In diffusion-controlled processes, the total flux per unit time is proportional to the total surface area available and the concentration gradient at the interface between quartz particle and viscous melt. The model assumes that a linear concentration gradient exists across the spherical shell of reaction products that invalidates the real situation where quartz particles are anything but perfect spheres.

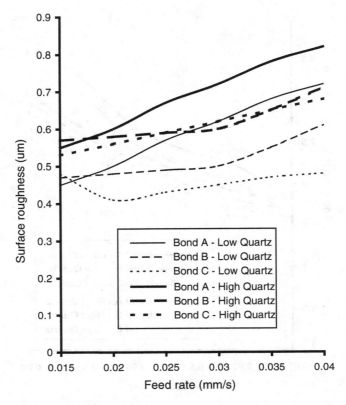

Fig. 2.39 Surface roughness as a function of feed rate for vitrified aluminum oxide grinding wheels with different types of abrasive and bond. Grinding conditions: wheel surface speed = 33 m/s; depth of cut = 0.25 mm; dress conditions = diamond blade traversed at 0.15 mm/rev at a depth of cut of 0.03 mm

The model does obey the parabolic law for diffusion around a sphere that can only apply when the sphere has a very low solubility in the solvent. However, this is not the case for quartz in most silicate systems. Also, the bonding between quartz particle and viscous glass melt is moving and not stationary. This implies that the concentration gradient is transient and not constant. The experimental data used to justify the accuracy of the dissolution model do not meet the conditions implicit in the model. The reactions between the three basic components of the bonding system, i.e., clay, feldspar, and quartz, cannot solely be described as diffusion-controlled dissolution of quartz in a liquid phase of constant composition and properties. It seems unlikely that the model can predict the mass fraction of quartz in any silicate system with great accuracy. However, over short soaking periods at the vitrification temperature, the results of the model compare well with experimental data.

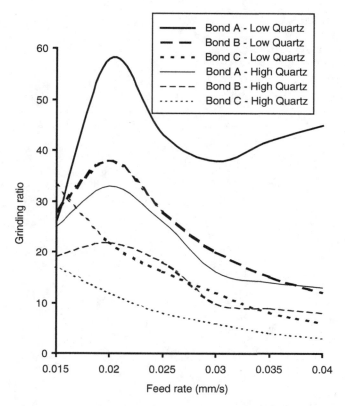

Fig. 2.40 Grinding ratio as a function of feed rate for vitrified aluminum oxide grinding wheels with different types of abrasive and bond. Grinding conditions: wheel surface speed=33 m/s; depth of cut=0.25 mm; dress conditions=diamond blade traversed at 0.15 mm/rev at a depth of cut of 0.03 mm

X-ray diffraction methods can be used to identify planes in vitrified bonding systems during and after heat treatment, which can be used to determine the correct bond formulations for grinding wheels used in specific applications. The experimental results show that bonds with high-quartz contents lead to much lower grinding ratios and surface roughness levels. This implies that X-ray methods can be used to increase the economic productivity of using such tools.

Conclusions

A review of the literature has provided a mathematical foundation that quantifies grinding wheel performance in terms of grinding parameters and wheel wear up to the burn boundary of a particular workpiece tested. The transformation of these fundamental factors presented in the form of performance diagrams shows optimum bond compositions and wheel grades for a particular workpiece–wheel combination under predetermined operating conditions.

A review of the literature has provided an insight into the causes and effects of wheel wear in terms of applied grinding loads, abrasive grain type, and abrasive strength. Relationships have also been formulated and tested that relate the wheel wear parameter, G-ratio, and the maximum tensile stress acting within the abrasive material. However, the effect of bond composition on wheel wear and performance has not been examined.

A review of the strength of clay-based materials thus provides a basis for studying the effect of bond composition on bond strength and wheel performance. Studies of the influence of workpiece material composition on grinding wheel wear also ensure that any study on grinding wheel performance should include factors that affect workpiece hardness and vitreous-bond strength.

The fired microstructure of all clay-based materials will depend on the structure of the raw materials used, the processing sequence, and the heat treatment schedule. The fired microstructure can contain:

1. Feldspar
2. Quartz and possibly cristobalite
3. A glass phase of variable composition
4. Cracks around quartz and cristobalite particles
5. Mullite-primary and secondary
6. Pores-open and/or closed, depending on the heat treatment, and fissure-like pores resulting from differential shrinkage

The complexity of the grinding wheel bonding system can be appreciated given the shape, size, amount, distribution, and orientation of constituent particles and how

M.J. Jackson and M.P. Hitchiner, *High Performance Grinding and Advanced Cutting Tools*, SpringerBriefs in Applied Sciences and Technology, DOI 10.1007/978-1-4614-3116-9, © Springer Science+Business Media New York 2013

these variables affect the bond's tribological properties such as wheel wear and grinding friction.

The behavior of abrasive cutting tools is dependent on the type of abrasive grain used, its heat treatment schedule, and its bond content. The vitrification behavior is dependent on certain processing variables and the bonding systems used for precision grinding wheels. It is evident that rutile needles have a significant effect on the mechanical properties and the behavior of conventional abrasive cutting tools. If titanium-doped aluminum oxide is used as the cutting medium then heat treatment cycles should be designed that prevent the growth of the needles into the bonding system. The development of strength during heat treatment depends upon the nature of the compounds formed at the interfacial layer between abrasive grain and bonding material. Further studies are required to understand the nature of this bonding with new abrasive grains and glass ceramic bond formulations.

Vitrified CBN grinding wheels tend to fracture at the bond bridge rather than at the interface between CBN and its glass bonding bridge. This implies that the relatively thin layer of boric oxide is stronger than the bonding bridge material. However, the choice of bonding composition may significantly influence the thickness of the interfacial layer that may become detrimental to the performance of the grinding wheel. This is particularly important if the failure mode during grinding is interfacial fracture between bond and abrasive grain.

The dissolution model derived by Jackson and Mills has been compared with experimental data using sintering and fusible vitrified bonding systems that are used extensively with high performance grinding wheels. The results predicted by the model compare well with the experimental results presented in this paper. However, over longer periods of isothermal vitrification, the model becomes less accurate due to the assumptions made in the dissolution model. The model may be of use when predicting the mass fraction of quartz using high temperature firing cycles that are characterized by short soaking periods. The models used are useful where bond bridge fracture wear is the dominant mode of grinding wheel wear.

Acknowledgments The authors thank Professor Givi Bockuchava for permission to use micrographs of abrasive materials from his extensive works on the wear of grinding wheels and grinding wheel structure. The authors also thanks the late Professor David Tabor, FRS, for advice and direction on the tribology of abrasive materials and the structure of diamond and cBN, whilst the principal author was a research fellow at the Cavendish Laboratory, University of Cambridge. The authors also thank Springer for allowing the authors to reproduce their own material that appeared in "Machining with Abrasives" published by Springer (License numbers: 2765660089893, 2765650671283, 2765650401549, 500633521. October 10th 2011 and 2767030346460, October 13th 2011). Published "With kind permission from Springer Science+Business Media: 'Machining with Abrasives', Edited by Mark J. Jackson and J. Paulo Davim, Springer Verlag, New York, November 2010, pp 1–423. ISBN 978-1-4419-7301-6".

References

1. De Pellegrin DV, Corbin ND, Baldoni G, Torrance AA (2002) The measurement and description of diamond particle shape in abrasion. Wear 253:1016–1025
2. De Pellegrin DV, Corbin ND, Baldoni G, Torrance AA (2008) Diamond particle shape: its measurement and influence in abrasive wear. Tribol Int 42(1):160–168
3. De Pellegrin DV, Stachowiak GW (2004) Evaluating the role of particle distribution and shape in two-body abrasion by statistical simulation. Tribol Int 37:255–270
4. Diashape—World Wide Web (2006) Diashape parameters, innovative sintering technologies. http://www.istag.ch/diamonds/parameters.html
5. Sysmex—World Wide Web (2006) FPIA-3000 Particle analyzer—parameters. Sysmex. http://particle.sysmex.co.jp/en/fpia/parameter.html
6. Verspui MA, van der Velden P, de With G, Slikkerveer PJ (1996) Angularity determination of abrasive powders. Wear 199:122–126
7. Hamblin MG, Stachowiak GW (1996) Description of abrasive shape and its relation to two body abrasion. Tribol Trans 39(4):803–810
8. Kaye BH (1984) Fractal description of fine particle systems. In: Beddow JC (ed) Particle characterization in technology, applications and microanalysis, vol 1. CRC Press, Boca Raton, FL
9. Swanson PA, Vetter AF (1984) The measurement of abrasive particle shape and its effect on wear. ASLE Trans 28(2):225–230
10. Leavers VF (2000) Use of the two-dimensional radon transform to generate a taxonomy of shape for the characterization of abrasive powder particles. IEEE Trans Pattern Anal Mach Intell 22(12):1411–1423
11. Roquefeuil F (2003) Abral: a new approach to precision grinding. Abrasive Magazine December:24–29
12. Komanduri R, Shaw MC (1975) Attritious wear of silicon carbide. ASME Paper # 75-WA/Prod-36
13. Kugemai N, Kiyoshi K (1984) Grinding of titanium with jet infusion of grinding fluid. In: Proceedings of the fifth International conference on titanium science and technology, 10–14 Sept 1984, Munich
14. Hirosaki K, Shintani K, Kato H, Asakura F, Matsuo K (2004) High speed machining of Bio-Titanium alloy with a binderless PcBN tool. JSME Int J C 47(1):14–20
15. Kumar KV (1990) Superabrasive grinding of titanium alloys. In: Fourth international grinding conference on SME, 9–11 Oct 1990, Detroit
16. Hagiwara S, Obikawa T, Usui E (1998) Edge fracture characteristics of abrasive grain. Trans ASME J Manuf Sci Eng 120:708–714
17. Ishikawa T, Kumar K (1991) Conditioning of vitrified bond superabrasive wheels. Superabrasives '91 SME MR91-172
18. Takagi J, Liu M (1996) Fracture characteristics of grain cutting edges of CBN in truing operation. J Mater Process Technol 62:397–402

M.J. Jackson and M.P. Hitchiner, *High Performance Grinding and Advanced Cutting Tools*, SpringerBriefs in Applied Sciences and Technology, DOI 10.1007/978-1-4614-3116-9, © Springer Science+Business Media New York 2013

19. Jakobuss M, Webster J (1996) Optimizing the truing and dressing of vitrified-bond CBN grinding wheels. Abrasives Magazine August/September:23
20. Fujimoto M, Ichida Y, Sato R, Morimoto Y (2006) Characterization of wheel surface topography in cBN grinding. JSME Int J C 19(1):106–113
21. Yonekura Y (1983) Effects of Tsukidashiryo of resinoid bonded borazon CBN wheels on grinding performance. Bull JSPE 17(2):113–118
22. Mindek R (1992) Improved rotary disc truing of hot-pressed CBN grinding wheels. MSc Thesis, University of Connecticut
23. Acheson EG (1893) Production of artificial crystalline carbonaceous material. US Patent 492,767
24. Cowles AH, Cowles EH (1885) US Patent 319945
25. Kern EL, Hamill DW, Deem HW, Sheets HD (1969) Mater Res Bull 4
26. H.F. Ueltz, 'Abrasive Grains - Past, Present, and Future', Proceedings of the International Grinding Conference - New Developments in Grinding, Keynote Paper, pp1–52, Edited by M.C. Shaw, Carnegie Press, Pittsburgh, Pennsylvania, USA, April 1972.
27. Nilsson O, Mehling H, Horn R, Fricke J, Hofmann R, Muller SG, Eckstein R, Hofmann D (1997) Determination of the thermal diffusivity and conductivity of monocrystalline silicon carbide. High temperatures-high pressure 29:73–79
28. Tymeson MM (1953) The Norton story. Norton Co, Worcester, MA
29. Wikipedia (2009) Bayer process. Online encyclopedia reading, 20 Nov 2009
30. Wolfe LA, Lunghofer EP (1998) New fused alumina production in South Korea and Australia. In: 13th Industrial minerals international congress, Kuala Lumpur, Malaysia, 26–29 Apr 1998
31. Lunghofer EP, Wolfe LA (2000) Fused brown alumina production in China. http://www.ceramicindustry.com/copyright/77d58fabca9c7010VgnVCM100000f932a8c0_...9/24/2009. Accessed 6 Aug 2000
32. Whiting Equipment Canada Inc. (2005) Metallurgical equipment. Commercial brochure
33. USGS Geological Survey (2009) Mineral commodity summaries. January 2009, United States Geological Survey, Reston, VA.
34. USGS Geological Survey (2008) Minerals yearbook, abrasives, Manufactured abrasives, United States Geological Survey, Reston, VA.
35. Wellborn WW (1991) Synthetic minerals—the foundation stone of modern abrasive tools. AES Magazine 31(1):6–13
36. Coes L (1971) Abrasives. Springer, New York, p 65
37. Polch (1956) US Patent 2769699, Nov 1956
38. Robie NP (1959) Abrasive material and method of making same. US Patent 2877104, 10 Mar 1959
39. Foot DG (1965) Mixture of fused alumina and fused granules in bonded abrasive articles. US Patent 3,175,8943/30
40. Marshall DW (1965) Fused alumina-zirconia abrasives. US Patent 3181939, 4 May 1965
41. Cichy P (1973) Apparatus for producing oxide refractory material having fine crystal structure. US Patent 3,726,6214/10
42. Richmond WQ, Cichy P (1975) Apparatus for producing oxide refractory material having fine crystal structure. US Patent 3,861,849, 21 Jan 1975
43. Richmond WQ, Cichy P (1975) Semi-continuous process for producing oxide refractory material having fine crystal structure. US Patent 3,928,515, 23 Nov 1975
44. Sekigawa H (1976) Process for manufacturing high strength Al_2O_3-ZRO_2 alloy grains. US Patent 3,977,132, 31 Aug 1976
45. Ilmaier B, Zeiringer H (1977) Method for producing alumina and alumina-zirconia abrasive material. US Patent 4,059,417, 22 Nov 1977
46. Cichy P (1977) Continuous process for producing oxide refractory material. US Patent 4,061,699, 6 Dec 1977
47. Ueltz HFG (1980) Fused alumina-zirconia abrasive material formed by an immersion method. US Patent 4,194,887, 25 Mar 1980
48. Richmond WQ (1983) Process for making oxide refractory material having fine crystal structure. US Patent 4,415,510, 15 Nov 1983
49. Richmond WQ (1984) Oxide refractory material having fine crystal structure and process and apparatus for making same. US Patent 4,439,895, 3 Apr 1984

50. Scott JJ (1976) Progressively or continuously cycle mold for forming and discharging a fine crystalline material. US Patent 3,993,119, 23 Nov 1976
51. Scott JJ (1978) Method of producing abrasive grits. US Patent 4,070,796, 31 Jan 1978
52. Rowse RA, Watson GR (1975) Zirconia-alumina abrasive grain and grinding tools. US Patent 3,891,408, 24 June 1975
53. Bange D, Wood B, Erickson D (2001) Developemnt and growth of sol gel abrasives grains. Abrasives Magazine June/July:24–30
54. Webster JA, Tricard M (2004) Innovations in abrasive products for precision grinding. In: CIRP innovations in abrasive products for precision grinding keynote STC G, 23 Aug 2004
55. Bauer R (1987) Process for production of alpha alumina bodies by sintering seeded boehmite made from alumina hydrates. US Statutory Invention Disclosure H000189, 6 Jan 1987
56. Leitheiser MA, Sowman HG (1982) Non-fused aluminum oxide-based abrasive mineral. US Patent 4,314,827, 9 Feb 1982
57. Wood WP, Monroe LD, Conwell SL (1989) Abrasive grits formed of ceramic containing oxides of aluminum and rare earth metal, method of making and products made therewith. US Patent 4,881,951, 21 Nov 1989
58. Bange D, Wood B, Erickson D (2001) Development and growth of sol gel abrasives grains. Abrasives Magazine June/July:24–30
59. Cottringer TE, van de Merwe RH, Bauer R (1986) Abrasive material and method for preparing the same. US Patent 4,623,364, 18 Nov 1986
60. Schwebel MG (1988) Process for durable sol-gel produced alumina-based ceramics, abrasive grain and abrasive products. US Patent 4,744,802, 17 May 1988
61. Rue CV (1985) Vitrified bonded grinding wheels containing sintered gel aluminous abrasive grits. US Patent 4,543,107, 24 Sept 1985
62. Pellow SW (1992) Process for the manufacture of filamentary abrasive particles. US Patent 5,090,968, 25 Feb 1992
63. DiCorletto J (2001) Innovations in abrasive products for precision grinding. In: Conference on precision grinding & finishing in the global economy—2001, Oak Brook, IL, 1–3 Oct 2001
64. Roffman R (2005) Natural industrial diamonds. Finer Point Magazine 3:26–30
65. Cockburn C (2002) Diamonds: the real story. National Geographic March:2–3
66. Trautman R, Griffin BJ, Scharf D (1998) Microdiamonds. Scientific American 8:82–87
67. USGS Geological Survey (2012) Historical statistics for mineral commodities in the United States, Data series 140, by T. Kelly and G. Matos, Reston, VA.
68. Ladd R (n.d.) Manufactured large single crystals. Finer Points—Wire Die Products and Applications 23–28
69. Bailey MW, Juchem HO (1993) The advantages of CBN grinding: low cutting forces and improved workpiece integrity. IDR Pt 3:83–89
70. Jackson MJ (2001) Vitrification heat treatment during the manufacture of corundum grinding wheels. J Manuf Proc 3:17–28
71. Krabacher EJ (1959) Trans ASME. J Eng Ind 81:187–200
72. Geopfert GJ, Williams JL (1959) The wear of abrasives in grinding. Mech Eng 81:69–73
73. Hahn RS (1962) On the nature of the grinding process. In: Proceedings of the third international machine tool design and research conference, Manchester, UK, Pergamon Press, UK, pp 129–154
74. Tarasov LP (1963) In: International research in production engineering—American Society of Mechanical Engineers (ASME), USA, Paper No. 21, p 196
75. Bhattacharyya SK, Grisbrook H, Moran H (1965) Analysis of grit fracture with changes in grinding conditions. Microtechnic 22:114–116
76. Malkin S, Cook NH (1971) The wear of grinding wheels. Trans ASME J Eng Ind 93:1120–1128
77. Jackson MJ (2002) Interfacial fracture of vitrified corundum. Trans North Am Manuf Res Inst Soc Manuf Eng 30:287–294
78. Decneut A, Snoeys R, Peters J (1970) Sonic testing of grinding wheels. Report MC 36, Nov 1970, Centre de Recerces Scientifiques et Techniques de L'industrie des Fabrications Metalliques, University of Louvain, Belgium
79. Winkler ER, Sarver JF, and Cutler IB (1966) Solid solution of titanium dioxide in aluminum oxide. J Am Ceram Soc 49:634–637

80. Walmsley JC, Lang AR (1988) In: Barrett C (ed) Advances in ultrahard materials application and technology, De Beers Industrial Diamond Division, UK, pp 61–75
81. Lundin ST (1959) Studies on triaxial whiteware bodies, Almqvist and Wiksell, Stockholm, Sweden
82. Malkin S, Cook NH (1971) The wear of grinding wheels—Part 1: attritious wear. Trans ASME J Eng Ind 93:1120–1128
83. Yoshikawa H (1963) Fracture wear of grinding wheels. International Research in Production Engineering—American Society of Manufacturing Engineers, Paper No. 23, p 209
84. Tarasov LP (1963) Grinding wheel wear grinding tool steels. International Research in Production Engineering—American Society of Manufacturing Engineers, Paper No. 21, p 196
85. Yoshikawa H, Sata T (1963) Study on wear of grinding wheels—Part 1: bond fracture in grinding wheels. Trans ASME J Eng Ind 85:39–43
86. Tsuwa H (1960) On the behaviour of abrasive grains in the grinding process. Technical Report of Osako University, Japan, vol 10, pp 733–743
87. Tsuwa H (1961) On the behaviour of abrasive grains in the grinding process—part 2. Technical Report of Osako University, Japan, vol 11, pp 287–298
88. Tsuwa H (1961) On the behaviour of abrasive grains in the grinding process—part 3. Technical Report of Osako University, Japan, vol 11, pp 299–309
89. Tanaka Y, Ikawa N (1962) Behaviour of abrasive grains on the diamond wheel. Technical Report of Osako University, Japan, vol 12, pp 345–354
90. Geopfert GJ, Williams JL (1959) The wear of abrasives in grinding. Mech Eng 81:69–73
91. Tsuwa H (1964) An investigation of grinding wheel cutting edges. Trans ASME J Eng Ind 86:371–382
92. Lal GK, Shaw MC (1972) Wear of single abrasive grains in fine grinding. In: Proceedings of the international grinding conference, Pittsburgh, USA, pp 107–126
93. Eiss NS (1967) Fracture of abrasive grains in grinding. Trans ASME J Eng Ind 89:463–470
94. Bhattacharyya SK, Grisbrook H, Moran H (1965) Analysis of grit fracture with changes in grinding conditions. Microtechnic 22:114–116
95. Mohun W (1962) Grinding with abrasive discs—parts 1, 2, and 3. Trans ASME J Eng Ind 84:431–482
96. Saito K, Kagiwada T (1974) Transient distribution of temperature and thermal stress in a grain due to a pulsating heat source. Bull Jpn Soc Precis Eng 8:125–126
97. Storch W, Ruf H, Scholze H (1984) Quartz particles contained in porcelain bodies. Berichte Deut Keram Ges 61:325
98. Binns E (1962) Refractory ceramics. Sci Ceram 1:315
99. Ford WF, White J (1951) The effect of heat on ceramics. Trans J Brit Ceram Soc 50:461
100. Kirchoff G, Pompe W, Bahr HA (1982) Acoustic emission study of cracks in porcelain bodies. J Mat Sci 17:2809
101. Jackson MJ, Mills B (1997) Dissolution of quartz in vitrified ceramic materials. J Mat Sci 32:5295–5304
102. Alexander IE, Klug HP (1948) X-ray diffraction procedures. Anal Chem 20:886
103. Khandelwal SK, Cook RL (1970) Effect of alumina additions on crystalline constituents and fired properties of electrical porcelain. Am Ceram Soc Bull 49:522–526
104. Jander W (1927) Reaktion im festen zustande bei hoheren temperaturen (Reactions in solids at high temperature). Z Anorg U Allgem Chem 163:1–30
105. Krause P, Keetman E (1936) Zur kenntnis der keramischen brennvorgange (On combustion processes in ceramics). Sprechsaal 69:45–47
106. Monshi A (1990) Investigation into the strength of whiteware bodies. Ph.D. Thesis, University of Sheffield, UK
107. Jackson MJ (1995) A study of vitreous-bonded abrasive materials. Ph.D. Thesis, Liverpool University, UK, December
108. Snoeys R, Leuven KU, Maris M, Wo NF, Peters J (1978) Thermally induced damage in grinding. Annals CIRP 27(1):571–581
109. Shaw MC (1991) Metal cutting principles. Clarendon, Oxford, p 410
110. United Grinding Technology (1995) Method of grinding titanium. WIPO Patent Application, WO/1995/003154
111. Campbell JD (1993) (UTC Hartford CT) Method of grinding titanium. US Patent, 5,203,122
112. Soo SL, Hood R, Lannette M, Aspinwall DK, Voice WE (2011) Creep feed grinding of burn reissistant titanium (BuRTi) using superabrasive wheels. Int J Adv Manuf Technol 53:1019–1026